AF736878

NANOTECHNOLOGY AND THE PUBLIC

Risk Perception and Risk Communication

PERSPECTIVES IN NANOTECHNOLOGY

Series Editor
Gabor L. Hornyak

Nanotechnology and the Public: Risk Perception and Risk Communication
Susanna Hornig Priest

Medical Nanotechnology and Nanomedicine
Harry F. Tibbals

Nanotechnology: The Business
Michael T. Burke

Nanotechnology: Ethics and Society
Deb Bennett-Woods

Nanotechnology Health and Environmental Risks
Jo Anne Shatkin

Nanotechnology: Legal Aspects
Patrick M. Boucher

<u>Forthcoming titles</u>

Nanotechnology and Global Sustainability
Donald Maclurcan

Nanotechnology Intellectual Property Rights: Research, Design, and Commercialization
Prabuddha Ganguli and Siddharth Jabade

NANOTECHNOLOGY AND THE PUBLIC

Risk Perception and Risk Communication

Susanna Hornig Priest

CRC Press
Taylor & Francis Group
Boca Raton London New York

CRC Press is an imprint of the
Taylor & Francis Group, an **informa** business

CRC Press
Taylor & Francis Group
6000 Broken Sound Parkway NW, Suite 300
Boca Raton, FL 33487-2742

First issued in hardback 2017

© 2012 by Taylor & Francis Group, LLC
CRC Press is an imprint of Taylor & Francis Group, an Informa business

No claim to original U.S. Government works

Version Date: 20110708

ISBN-13: 978-1-4398-2683-6 (pbk)
ISBN-13: 978-1-138-45351-7 (hbk)

This book contains information obtained from authentic and highly regarded sources. Reasonable efforts have been made to publish reliable data and information, but the author and publisher cannot assume responsibility for the validity of all materials or the consequences of their use. The authors and publishers have attempted to trace the copyright holders of all material reproduced in this publication and apologize to copyright holders if permission to publish in this form has not been obtained. If any copyright material has not been acknowledged please write and let us know so we may rectify in any future reprint.

Except as permitted under U.S. Copyright Law, no part of this book may be reprinted, reproduced, transmitted, or utilized in any form by any electronic, mechanical, or other means, now known or hereafter invented, including photocopying, microfilming, and recording, or in any information storage or retrieval system, without written permission from the publishers.

For permission to photocopy or use material electronically from this work, please access www.copyright.com (http://www.copyright.com/) or contact the Copyright Clearance Center, Inc. (CCC), 222 Rosewood Drive, Danvers, MA 01923, 978-750-8400. CCC is a not-for-profit organization that provides licenses and registration for a variety of users. For organizations that have been granted a photocopy license by the CCC, a separate system of payment has been arranged.

Trademark Notice: Product or corporate names may be trademarks or registered trademarks, and are used only for identification and explanation without intent to infringe.

Visit the Taylor & Francis Web site at
http://www.taylorandfrancis.com

and the CRC Press Web site at
http://www.crcpress.com

Contents

Perspectives Contents vii
Acknowledgments ix
Perspectives Contributors xi

1 **Risk Communication in a Democratic Society** 1
Risk and Technology 3
Technology and Society 6
References 19

2 **Introducing Nanotechnology to the Public** 21
Imagining the Nanoscale 22
Emerging Public Perception 26
Risk Communication for Twenty-First Century Democracies 29
References 37

3 **Risk Communication in Theory and Practice** 39
Risk Communication Challenges 40
The Goals of Risk Communication 44
Social Theories of Risk 46
References 55

4 **Public Opinion, Public Perception, and Public Understanding** 57
The GM Food Story Revisited 58
Opinion Studies and Their Implications 61
Implications for Risk Communication Research and Practice 65
Nanotechnology and "Cultural Resonance" 68
References 78

5 **What Do People Want from Technology?** 81
Technological Literacy and Democracy 82
The Challenges of Risk Society 85
Nanotechnology, Risk, and Society 89
References 112

6 **Audiences, Stakeholders, Cultures, and Nanotechnology Risk** 113
The Role of Advocacy in Democracy 116
Social, Cultural, and Psychological Influences 120
Persuasion Research 124
References 139

7 **Disseminating Information about New Technologies** ... 143
Media's Role in Risk Societies ... 146
Media Effects Theory in the Internet Age ... 150
Museums and Science Centers ... 155
References ... 166

8 **Lessons and Future Challenges** ... 167

Index ... 185

Perspectives Contents

Chapter 1 Perspective: Where Nano Came From 9
Matthew N. Eisler

Chapter 2 Perspective: Professional Principles of Risk Communication 33
Joye Gordon

Chapter 3 Perspective: The National Research Council's 1989 Recommendations for Improving Risk Communication: Risk Communication as an Interactive Process 51
Thomas B. Lane, Jr.

Chapter 4 Perspective: Public Attitudes Toward Nanotechnology 71
Dietram A. Scheufele

Chapter 5 Perspective Part 1: Lessons Learned from the First U.S. National Citizens' Technology Conference 93
Michael D. Cobb

Chapter 5 Perspective Part 2: Moving Forward with Citizen Deliberation: Lessons and Inspiration from the National Citizens' Technology Forum 103
Jason Delborne and Jen Schneider

Chapter 6 Perspective: Nanotechnology Governance and Publics: Making Connections in Policy 126
Jennifer Kuzma

Chapter 6 Perspective: Public Policies and Public Perceptions of Nanotechnology in the United States and in Europe 132
Esther Ruiz Ben

Chapter 7 Perspective: What Have the Mass Media Been Reporting on Nanotechnology Risks? 157
Sharon M. Friedman and Brenda P. Egolf

Chapter 8 Perspective: The Ethics of Risk Communication for Nanotechnology 171
Paul B. Thompson

Acknowledgments

The editorial assistance of graduate students Thomas Lane and Gabriel Reuben Young at the University of Nevada, Las Vegas, is gratefully recognized. Without their organizational help, this project could not have been completed. Portions of the research conducted by Priest and discussed here received financial support from the National Science Foundation through grants #0758195, #0531160, and #0523433. Any opinions, findings, and conclusions or recommendations expressed in this publication are those of the respective authors and do not necessarily reflect the views of the National Science Foundation or any other individual or organization.

Perspectives Contributors

Michael D. Cobb
College of Humanities and Social Sciences
North Carolina State University
Raleigh, North Carolina

Jason Delborne
Liberal Arts and International Studies
Colorado School of Mines
Golden, Colorado

Brenda P. Egolf
Center for Social Research
Lehigh University
Bethlehem, Pennsylvania

Matthew N. Eisler
Center for Nanotechnology in Society
University of California, Santa Barbara
Santa Barbara, California

Sharon M. Friedman
Department of Journalism and Communication
Lehigh University
Bethlehem, Pennsylvania

Joye Gordon
School of Journalism and Mass Communication
Kansas State University
Manhattan, Kansas

Jennifer Kuzma
Humphrey Institute of Public Affairs Science
University of Minnesota
St. Paul, Minnesota

Thomas B. Lane, Jr.
University of Nevada, Las Vegas
Las Vegas, Nevada

Esther Ruiz Ben
Department of Sociology
University of Essex
Essex, United Kingdom

Dietram A. Scheufele
Department of Life Sciences Communication
University of Wisconsin-Madison
Madison, Wisconsin

Jen Schneider
Liberal Arts and International Studies
Colorado School of Mines
Golden, Colorado

Paul B. Thompson
W.K. Kellogg Chair in Agricultural, Food, and Community Ethics
Michigan State University
East Lansing, Michigan

1

Risk Communication in a Democratic Society

A defining characteristic of being human is making and using technology; a defining characteristic of human society today is a collective interest in managing technology, along with both its benefits and its risks. We could hardly survive, or lead our present lives, without technology. Technology has extended our life span and provided novel solutions to fulfilling our basic needs, from shelter to transportation, energy, and beyond. Technology is indispensable to modern agriculture. Yet many—perhaps most—technologies provide risks as well as benefits. As technology has become more complex, and its societal implications more clearly far reaching, some scholars have begun to characterize today's highly developed societies as "risk societies" (Beck 1992) in which the management of technological risk has become the central concern around which these societies are organized. Such analyses are controversial. Certainly, however, modern technology is both blamed as a root cause and celebrated as a potential solution for many vital contemporary policy issues, ranging from the problem of environmental degradation to the challenge of expanding health-care options, from the need to feed a burgeoning world population to the necessity of developing energy-efficient transportation. Technology (especially energy-consumptive technology) is largely responsible for the looming threat of CO_2-induced climate change, as well as offers potential remedies such as alternative energy production and alternative forms of transportation that contribute less to the average individual's *carbon footprint,* alongside more radical and controversial solutions such as geoengineering. Critics refer to our hopes that technology can always be called on to solve the problems, in turn, of technology as the *technological fix,* an assumption that seems ubiquitous in modern times.

Nanotechnology, most often defined as "the understanding and control of matter at dimensions between approximately 1 and 100 nanometers, where unique phenomena enable novel applications" ("What is nanotechnology?" n.d.), has been linked directly to all of these developments and more, offering the potential of advancing our options in energy, environment, medicine, and agriculture. It is well beyond the scope of this book to discuss in detail the scientific and engineering aspects of nanotechnology. However, it is important for our purposes to recognize that nanotechnology is not a "thing." Rather, it is a class of things—a broad group of technologies, both those presently available and those imagined for the future—that are being brought about by advances in our ability to see and manipulate matter at what is referred to as the *nanoscale*. These advances have been brought about primarily by advances in microscopy, including the development of atomic force and scanning tunneling microscopes that allow scientists to probe the properties of nanoscale matter. Material can have unique properties when the particles are this small. These properties have been harnessed in many applications and are expected to be utilized in many more.

For example, nanotechnology-based sensors might help monitor the environment, while other developments in nanotechnology might help more directly control pollutants; nanotechnology-based catalysts might make combustion engines vastly more efficient; nanotechnology in food and agriculture might help agricultural chemicals reach their targets more efficiently, with less chemical waste and cost to the environment; nano-encapsulated drugs might destroy human tumors without damaging surrounding tissues; and nanotechnology-based food packaging might be able to identify product spoilage and thus prevent food poisoning. Already, nanotechnology in electronics has helped move us farther along the path to ever-smaller components and more efficient information processing, helping engineers to imagine entirely new approaches to computer architecture. Nanotechnology in materials science has helped make sporting equipment like stronger skis and cosmetics that are better able to penetrate the skin. Nanoparticles of titanium dioxide or zinc oxide turn sunscreen from opaque white to transparent. Washing machines can inject nano-size silver particles into the wash water, a feature intended to make clothes smell fresher (by killing bacteria), and clothing that incorporates nanomaterials is designed to release stains and resist wrinkles.

All this is only the beginning. A database created by the Woodrow Wilson Center's Project on Emerging Nanotechnology already lists over 1000 products in which nanotechnology is used ("Nanotechnology consumer products inventory," 2010). At the farthest edge of the horizon, futurists imagine that nanotechnology might be used to augment human memory or otherwise enhance the capacity of the human body. Only time will tell what the future of nanotechnology will actually bring; no doubt some of the developments we hope for will never come to pass, but others will, and yet others we have not even begun to contemplate will follow as well. Much optimism surrounds the use of nanotechnology in health and medicine; envisioned applications include cancer treatments that can target particular cells and tiny cameras that can look for abnormalities in the digestive system. Nanoparticles (in the form of *quantum dots*) are used in the manufacture of solar panels; carbon nanotubes (which can be absorbed by the human body, where they may contribute to the development of mesothelioma just as asbestos does) have a variety of applications in electronics, as well as add strength to a variety of manufactured materials.

Risk and Technology

Yet, like all technology, nanotechnology also poses risks. Modern societies have learned several painful lessons from observing the trajectories of past technologies. The first is that almost all major technological developments—nuclear power, chemical pesticides and fertilizers, gene therapy, modern medicines, food additives, the automobile, and the airplane, to name only a very few of the most common examples—have risks. This often includes risks to the environment, risks to human health, and risks to other species. Global climate change, clearly among the most compelling of contemporary risks because of its all-encompassing nature and the limitations of available solutions, is a tragic result of our reliance on polluting technologies (at least in part, if not in its entirety). Nanotechnology, like earlier technologies, undeniably carries risks as well as benefits, and many of these risks are unknown. If nanosilver is released into wash water, what is its fate in, and effect on, the environment? If cosmetics and sunscreen can penetrate the skin, what is their fate in, and effect on, the body? For the most part, such questions do not yet have answers. Risks associated with nanotechnology and products using nanotechnology remain under investigation, and our understanding of these risks is not yet comprehensive. Government regulators struggle to keep up with these emerging risks and to develop regulatory systems that are capable of responding to them in a timely way.

Technologies also raise other questions and concerns. These include concerns over ethics (as in the case of stem cell research), access to benefits (as in the case of expensive new medical treatments), exposure to risks (as illustrated by issues of environmental justice with respect to the locations of toxic dumps and pollution-generating industries), and economic impacts (both good and bad, because, while large-scale technological change can create jobs, it can also destroy them). Particular concerns surround the exposure of researchers and factory workers to new nanomaterials, entailing largely unregulated and often poorly understood risks. Ordinary citizens may not understand the technical details of how nanotechnology works, but they often raise reasonable questions about who will benefit, and who may be differentially exposed to the associated risks. At the same time, however, citizen reactions to nanotechnology and its applications have not engendered the level of negative response associated with some earlier technologies that have become sources of ongoing controversy, such as nuclear power or agricultural biotechnology.

Agricultural biotechnology, which brought us foods and other crops genetically modified to resist insect damage, tolerate chemical weed killers, last longer in storage, survive frosts in the field, and even grow vaccines, among other potential benefits, were introduced largely without public discussion, with little demonstrated consumer benefit, and in the absence of comprehensive data on risks. They became (and to some extent remain) extremely divisive, perhaps in part because many people were simply unaware of their deployment. This history suggests we should give more attention to the best ways to encourage public discussions about risks, as well as public consultation[1] regarding whether the benefits of particular technologies are worth those risks. From an industry perspective, money spent on developing products and processes that people reject (rightly or wrongly) can be money wasted. Experience with the emergence of public concerns over agricultural biotechnology in particular, which generated substantial resistance to adoption of these technologies in both Europe and the United States, has suggested to those developing the next generation of technology—that is, nanotechnology—that they should take public perception seriously, even where they may disagree about the seriousness of particular risks. As a result, social scientists generally and risk communication specialists in particular have been enlisted in the effort to anticipate and respond to possible public concerns about this new class of technology.

Experience also suggests that laypeople and scientific experts may understand risk quite differently. Many engineers and scientists are trained to define risk narrowly, as the probability of being subjected to a reasonably well-defined, identifiable, measurable, physical hazard. But "ordinary" (that is, nonexpert) people, on the other hand, are also concerned about the broader societal implications mentioned above, what we might think of as *social risks* involving ethics, justice, and economics, as opposed to specific risks of physical harm. Our *risk societies* are gradually learning how to cope with the physical risks of technology but often shy away from active management of these social risks. Both physical and social risks tend to come into play when technologies become controversial. They are often intertwined, even though each individual technology represents a unique constellation of public concerns. It is often difficult to separate the narrow class of physical risks, with which formal risk assessment has traditionally been concerned, from other concerns over ethics and justice. It is also difficult to separate physical risk from even less tangible concerns

[1] In this context, the term "public consultation" means (literally) to consult the public, implying not only assessing public opinion through surveys, focus groups, and other structured discussion, but also taking public opinion into account in making policy decisions.

reflected in more purely emotional reactions that often seem to have a cultural foundation but are more difficult to articulate. This is not to assert that emotional reactions to a given risk are necessarily irrational or without any basis in fact, but to emphasize that our experience of risks often entails both emotional and cognitive elements.

For example, cloning may generate an initial reluctance that we might attribute to ethical concerns; this reluctance seems to be strongest when we consider human cloning, weaker but still present when we consider animal cloning, and weakest when considering the cloning of bacteria (as might be utilized in drug production). The possibility that a cloned being—especially a person or, for some, an animal—might have defects manifesting as health problems, which represent a risk of physical harm, becomes intertwined with concerns over the broader ethics of cloning, which to some people simply seems frightening, unnatural, or inherently wrong in ways they may not be able to explain. The idea that cloned animals might enter the food supply makes some people feel uneasy as well, if only because it just does not "feel" natural. Some of the concern over human cloning has centered on the psychological effects on someone who finds out he or she has been created to replace someone in the family who has died, or to produce a transplantable replacement organ for another family member. Such scenarios may concern us both emotionally and rationally.

Yet, each new technology is unique. So far, nanotechnology has produced relatively weak public reactions and little protest, given its transformative potential. In retrospect, this underscores the possibility that deep cultural factors may have generated public reactions to genetically modified food, human and animal cloning, and other manifestations of contemporary biotechnology, factors that appear (for the most part) absent from initial public reaction to nanotechnology. However, as nanotechnology is further developed, particularly in the form of *nanobiotechnology* (that is, various combinations of nanotechnology and biotechnology designed for use in agriculture or medicine), deeply rooted concerns about bad outcomes may well arise for it as well. Some aspects of nanotechnology could possibly make some individuals feel uneasy for a variety of reasons, including aspects involving exposure to various identifiable physical risks, elements of "social risk" involving distribution of benefits or other definable ethical challenges, and reactions that seem to result simply from a technology's striking people as somehow inherently wrong or unnatural.

Whatever the reasons, the apparently weak levels of public concern that surrounds the adoption of nanotechnology stand in contrast to reactions to various forms of biotechnology that have engendered serious objections around the globe, despite rather limited demonstrable evidence of harm (except in isolated cases, e.g., via gene therapy). Nanotechnology has resulted in little negative public reaction, despite emerging evidence of risk to health and environment for at least some of its forms. This observation is not to say that popular objections to biotechnology have necessarily been unfounded. It does raise the intriguing question of why one set of technologies should create such serious popular concern, while another is met with broad public acceptance, despite expert opinion that the latter might present important risks alongside its benefits, whereas the risks of the former may be more remote. By better understanding how people react to new technologies—that is, how underlying concerns about technology arise and persist—efforts to educate, to provide opportunities for public discussion, and to manage and regulate technology in a socially appropriate manner can be improved. The efforts of risk communication specialists are integral to such efforts.

Technology and Society

Philosophers and ethicists tend to distinguish between ethical issues associated with outcomes, including physical, social, or psychological harm. An example is the assertion that human cloning might be wrong because it has bad effects on individual organisms (including people). There are also ethical issues associated with the violation of basic rights or other ethical principles, such as the idea that it may simply be inherently wrong to clone people because the worth and value of individual humans lies partly in their dignity and their uniqueness, which cloning is seen as desecrating. The first type of issue analysis, concerned with outcomes, is called *consequential* or *utilitarian*, and the second may be referred to as a *rights-based* or *rules-based* perspective. Similarly, a particular form of medical technology such as gene therapy that made possible a permanent alteration to the human body might be evaluated in terms of its potential benefits and harms alone, or in terms of an argument that it is (or is not) inherently wrong to alter an individual's genetic makeup. (Some medical ethicists may distinguish between alterations intended to cure an illness and alterations intended to "improve" inherited traits that fall within the normal range.) One contemporary prediction is that in the future, new forms of nanotechnology now under development may facilitate gene therapy.

These two categories of ethical analysis (utilitarian and rights based) are not always distinct in practice, however. In addition, what social psychologists call cognitive reactions (the results of thought, reasoning, and rational analysis) and emotional or affective reactions (less analytical reactions involving feelings that we cannot necessarily fully explain in rational terms) are not clearly separated but often co-occur, with the boundaries between one and the other not always obvious. We can imagine a reasonable person arguing (whether or not we agree with him or her) that it is inherently wrong to clone human beings (a rules-based perspective), in part *because* we might have a particular psychological reaction to the individual (a utilitarian concern, although one involving potential emotional effects rather than physical harm). We might be concerned about human cloning or made uneasy by the prospect of an actual human clone because of what reason might tell us about the loss of respect for human individuality that cloning might entail. Or, we might have a more purely emotional reaction that does not seem to result from conscious cognitive processes but from a conviction that the act is simply wrong. Of course, technologies we choose to develop are generally intended to have positive outcomes, not just negative ones. This is why human beings seek to develop technologies to begin with (acknowledging that we can undoubtedly make mistakes, that some technologies might be made for inherently destructive purposes such as fighting a war, and that others can carry risks we did not fully foresee and may fail to control, as in nuclear power generation).

We use technology in hopes of meeting our most basic needs for food, shelter, and clothing better and more efficiently, as well as of making life more interesting, enjoyable, and entertaining. Even such a controversial technology as human cloning can certainly have positive uses. Perhaps the life of an ill sibling might be spared if an "expendable" organ such as a single kidney can be donated from a newly cloned individual to the one who is ill, a procedure that has the best chance of success if the organ comes from someone who is genetically identical. Perhaps this can be accomplished without disrespecting the inherent human worth of the clone. Yet questions remain about whether this kind of radical solution can (or should) be made available to everyone, to only those with the most life-threatening illnesses, or to only a lucky few with the resources to make such a choice possible. Nanotechnology does not yet seem to present the same kinds of concerns in the minds of most of us, but ultimately some of its applications may be equally controversial.

Despite all the controversy that can erupt over some technologies, people in the United States and other highly developed societies tend to embrace technology more than they reject it, in recognition of the contributions technology makes to our quality of life. This tendency also has deep cultural roots. But, people also recognize, often on the basis of our collective experience, that there are risks and ethical conundrums associated with many, if not most, technologies. Even though nanotechnology has not become broadly controversial, at least not yet, both experts and nonexperts are beginning to recognize that this class of technologies carries with it environmental and health risks, as well as potential ethical concerns, alongside its many benefits. Almost all broad classes of technology present negatives as well as positives.

Nearly all technologies, then, present a mix of benefits and physical and social risks, including the risk that either the technology's bad effects or its good ones, or both, will be distributed unfairly, and the possibility that other controversial dimensions may arise. As technologies first emerge, there are often more uncertainties than answers as to what benefits will actually be realized by most people and what foreseen and unforeseen risks will emerge—nanotechnology is no exception. In addition, questions such as what is an unacceptable risk, what is unfair, and what benefits are most worth having are not matters of science or engineering, narrowly defined. These questions are determined by human values. The adoption of new technology is generally a choice that is heavily laden with such values and that can therefore become controversial in any given case. Once adopted, technologies may prove difficult to give up (e.g., consider the automobile or the cell phone). But initially, adoption or nonadoption is in most respects a collective, if not always an individual, choice. There is usually no simple, single, scientifically correct answer available as to how risky a technology—let alone a broad class of technologies such as nanotechnology—might prove to be, especially when the social as well as the physical aspects of risk are considered and when both ethical quandaries and emotional reactions are factored in. Hoped-for benefits may or may not materialize as expected. Technology thus presents us with powerful potentialities and great uncertainty.

Sometimes people in the engineering and science communities become concerned that society has become too worried about risks and not aware enough of valuable benefits. People can also become too carried away with the potential benefits and overlook the risks, as in the early years of automobile use before the need for seatbelts, roll bars, better roads, and other safety features was widely recognized, or in the early period of pesticide use when the unintended effects on various living species, including humans, and on ecosystems were not yet apparent. When a technology is newly emerging, there is usually no right answer as to how much benefit is likely and whether it is worth the associated risk. The field of risk analysis, and its application in risk management, has emerged to try to do a better job of foresight in these areas. The field of risk communication has arisen in an attempt to provide better ways to communicate with broader publics about potential risks.[2]

Experience with previous technologies has suggested that the right thing to do, both ethically, as the right way to practice democracy, and practically, as the strategically best way to manage new technology, is to bring a

[2] The word publics is often preferable to phrases such as general public or mass public, because social scientists, including communication researchers, generally recognize that there really is no such thing as a uniform general or mass public, especially in modern democracies, which tend to be pluralistic. However, because using publics in the plural may be confusing to some readers, this text generally uses the phrase broader publics instead, while trying to avoid the misleading term general public or the mass public concept.

broad spectrum of the members of society on board as partners in thinking through the wisdom of new technologies from the beginning. This idea is sometimes referred to as public engagement or, more particularly, *upstream engagement*, meaning the engagement takes place before final decisions about technologies are made and before major controversy has arisen. The field of risk communication has been greatly influenced in recent years by this concept, and the development of nanotechnology has taken place in an environment in which considerable investment has been made in trying to improve our capacity to consult with broader publics outside of the scientific, engineering, and policy communities from the beginning.

It is now widely recognized that one of the things that exacerbated public concern about many previous technologies, agricultural biotechnology in particular, was the failure to engage in broad public discussion about the adoption decision. This kind of discussion is not intended to just educate people about science and technology or to make them feel better because they believe they have more of a say in making an important choice, but to allow them to participate in a meaningful way in developing decisions involving the wise use of technology. As will be discussed in later sections of this book, public engagement is not a panacea; its purpose is public empowerment through informed consultation, and it will not always prevent technology from becoming controversial. However, it is hoped that, at a minimum, public engagement will provide an early warning system of public concerns, allowing managers and regulators to consider those concerns. And, ideally, it should put democracy on a sounder footing by giving broader publics a say in how technologies are managed and even, ultimately, whether or not they are adopted. This book is about how public perception is emerging for nanotechnology, and how our knowledge of public preferences and public concerns can improve the practice of risk communication for nanotechnology.

PERSPECTIVE: WHERE NANO CAME FROM

Matthew N. Eisler

The etymology of the term *nano* and the family of compound neologisms it prefixes, as well as the sorts of material practices and social relations they signify, are shrouded in obscurity and uncertainty. Perhaps no other family of words in the English language is as misunderstood, exalted, parsed, or excoriated. Derived from the Greek word for "dwarf," *nano* is the metric expression of a factor of 10^{-9} or (0.000000001). As a unit of scale signifying one-billionth of a meter, the nanometer had become commonly used in many fields of science, engineering, and technology by the first decade of the twenty-first century to measure natural objects and fabricated substances.

But nano is not simply a measurement of scale. Over the last two or three decades, it has become a marker of progress in science and technology and a bellwether of changing ideas of the nature of innovation. Many nano-prefixed compounds were invented in this period to denote actual capabilities in imaging, manipulating, and understanding matter on this scale. In addition to *nanometer*, these included *nanoscale*, *nanoparticle*, *nanophase material*, *nanomaterial*, and *nanoscience*. Perhaps the best-known, yet least well defined and most problematic of these is *nanotechnology*. It and other nano-prefixed terms are frequently used interchangeably. For example, the architects of the National Nanotechnology Initiative (NNI) define *nanotechnology* as the understanding and control of matter at the scale of 1 to 100 nanometers, but then note somewhat tautologically that this may encompass nanoscale science, engineering, and technology, or NSET (Subcommittee on Nanoscale Science, Engineering, and Technology, Committee on Technology, National Science and Technology Council, 2010).

In fact, there are important and often subtle differences between the nano terms. Unlike its relatives, which tend to be used in specific contexts, *nanotechnology*, a word that describes no single device, material practice, or discipline, has acquired a variety of connotations thanks to its appropriation by visionary writers and U.S. federal science policy makers.

Disaggregating the real and the ideal from the multiple senses of nano reveals a complex history. Both the nano lexicon and the term nanotechnology originated in foreign science and technology milieux. But in the 1990s, in response to sustained Congressional scutiny of the economic value of U.S. federally-supported academic science, nanotechnology was adopted for political purposes by American science and science policy figures in expressing a revolutionary form of applied science with industrial relevance.

As multiple meanings of nano were shaped in a succession of social contexts over time, they accreted, complicating their definition. Ann Johnson noted that social and physical scientists often take the concept of nanotechnology for granted, failing to clarify what it is and who practices it (Johnson 2009, 144). Recent research suggests that nanotechnology is less a distinct, coherent body of material science or engineering practices than an umbrella term signifying iterative developments over the last quarter century, especially in the field of materials microfabrication (Mody 2004, 2009; Schummer 2006; Choi and Mody, 2009; Gallo 2009; Johnson 2009; Nordmann 2009). So, too, have scholars deconstructed nanotechnology's creation myths (Toumey 2005). One of the most prevalent is that its intellectual provenance can be directly traced to a speech given by the physicist Richard Feynman on December 29, 1959, in which he discussed manipulating matter at the atomic and molecular scales. As the historian W. Patrick McCray notes, Feynman never did any research along these lines (2005, 181).

But in later decades following this speech, various groups would cite the storied physicist, a pioneer in the field of quantum electrodynamics and winner of the Nobel Prize in Physics in 1965, as the inspiration for certain fields of materials research. Feynman's speech remained largely unknown until it was prominently cited in the November 1979 issue of *Physics Today*. This occurred not in reference to nanotechnology, a word that did not appear in the article, but to what the authors called *microscience*. James A. Krumhansl and Yoh-Han Pao defined this as a bundle of practices including methods of characterization and precise top-down microfabrication of structures that drew from "physical and engineering scientific knowledge" in solid-state physics, chemistry, materials science, and electric systems science. Reflecting the conventions of the day, Krumhansl and Pao expressed this work not in nanometers but in angstrom (Å), (0.01 of a nanometer), as had Feynman. In 1979, microscience occurred on scale of 100 to 1000Å (10 to 100 nm) and was within striking distance of 10 to 100 Å (1 to 10 nm) (Krumhansl and Pao 1979, 29).

Two years later, K. Eric Drexler, an aerospace engineer with a taste for sweeping visionary schemes, cited Feynman in a paper published in the *Proceedings of the National Academy of Sciences* (1981). Inspired by contemporary developments in computing and, above all, genetic engineering, the breaking up of and reconstruction of DNA, Drexler mulled the possibility of developing "molecular engineering," an imagined form of "microtechnology." He envisioned this as a method of manufacturing based on the biological processes of self-assembling protein molecules but utilizing more robust materials. Computer simulation, he claimed, suggested that such a technology was, in principle, capable of building atomically precise structures from the bottom-up. Subsequent advances in biological and characterization technology, including the invention of the scanning tunneling microscope by Gerd Binnig and Heinrich Rohrer in 1981, lent credence to Drexler's ideas (Regis 1995). Several years later, Drexler expounded on them at length in *Engines of Creation* (1986), a book-length monograph intended for popular consumption. In this work, Drexler supplanted microtechnology with nanotechnology as his chief synonym for molecular engineering.

The word *nanotechnology* was struck in 1974 by the Tokyo Science University professor Norio Taniguchi to describe materials-processing technologies capable of achieving ultrafine finishes of one nanometer. Reviewing a variety of existing devices for their potential for broader application in the industrial manufacturing of machinery, electronics, and opto-electronics, Taniguchi identified molecular beam epitaxy and ion sputtering, the former creating crystalline films through the deposition of atoms and the latter using the kenetic energy of inert gas ions to smooth surfaces by blasting atoms from them, as the best candidates for future development owing to their inherent precision and control (Taniguchi 1974, 18–23).

However, nanotechnoogy did not then become a synonym for these devices. The word remained obscure until adapted and popularized by Drexler in an utterly different sense. In addition to employing *nanotechnology* as a synonym for *molecular engineering*, Drexler used it as the basis of a historical philosophy outlined in *Engines of Creation*, one that conflated biological evolution with social acquisition, or, more precisely, technological innovation. Citing the Feynman speech, Drexler saw technological progress as sociobiological in essence, adapting the idea of the meme, an invention of Richard Dawkins. Technological progress as a sort of memetic evolution of molecular assemblers, wrote Drexler, could one day manifest as a sort of super pandemic devasating the biosphere in a tide of self-replicating assemblers, the infamous "grey goo," unless its development was mediated by a special class of highly trained experts (Drexler 1986, 35–37, 171–190).

In one sense, Drexlerian nanotechnology was a kind of metaphysics (Regis, 1995; Nordmann 2009). But *Engines of Creation* also outlined an idealized political economy of science, engineering, and innovation, one that justified technocracy and free market ideology and channeled deep-seated beliefs in American culture of the place of the individual in science and technology and the role of science and technology as social practices. Drexler's promise of a transdisciplinary field of virtually unlimited potential paralleled larger hopes for collaborative research and development in a period when Congress was promulgating a series of laws aimed at bolstering the creative innovative potential of small business, government-owned and -operated laboratories, and, above all, the American university, by establishing private property rights for ideas produced with public money. In the future, wrote Drexler, social policy would be irrelevant thanks to nanotechnology's incredible fecundity. Assemblers would relieve humans of the divisive task of redistributing resources from a limited terrestrial pie by creating vast new wealth from the limitless cornucopia of the cosmos (Drexler 1986, 93–98).

These ideas would influence U.S. federal science policy in the 1990s. Nano-prefixed science terminology, on the other hand, appeared much earlier around 1981, thanks in good measure to the German materials researcher Herbert Gleiter. He condensed metal vapors to produce a new class of what he later called "microcrystalline" materials, a special state of gas-like solid matter different from the glassy and crystalline state that could be alloyed on the "nanometer scale" (Nordmann 2009, 129–130).

Gleiter's recognition of the size-dependent properties in nanoscale materials, his use of nano-prefixed language, and his call for a research program in this field would, argues Nordmann, qualify him as a plausible "father of nanotechnology." That Gleiter was not so lauded, argues Nordmann, was the result of theoretical premises that clashed with the rhetoric and metaphysics of nanotechnology prevalent by the early 1990s. Thus, the theory-based expression "nanophase materials," coined in the late 1980s by the materials researcher Richard Siegel to describe the sorts of materials Gleiter investigated, fell out of use by the early 1990s. Nordmann holds that the notion of substances existing in a state of "gas-like disorder" was at odds with the ideas of precise characterization and control, and fabricated nanoscale devices that were increasingly associated with nanotechnology by this period. When *nanophase* was subsequently employed, it was used in senses divorced from its theoretical roots (Nordmann 2009, 137–140).

As Mody notes, there was no inherent logic in adopting nano-prefixed terminology to describe phenomena at that scale. Within microfabrication communities, existing nomenclature was perfectly adequate. Researchers altered their language, he suggests, to express technological progress and, hence, accrue symbolic and real capital at a time when the structure and political economy of U.S. basic science in industry and academe was in flux. Encouraged by instrument manufacturers eager to promote scanning tunneling/atomic force microscopy from the mid-1980s into the early 1990s, academic microfabrication culture conjugated terms like *nanoengineering, nanofabrication,* and *nanostructure fabrication* as indices of their ability to characterize and manipulate matter on the nanoscale. But they also created these new compound terms as a means of broadening disciplinary participation as industry's contribution to academic science waned, especially at eastern universities (Mody 2011).

Nanotechnology, on the other hand, held connotations distinct from these terms thanks largely to Drexler's activities as a popularizer, which included congressional testimony in 1992. The word began to gain currency in U.S. science policy circles around this time, argues McCray, because its utopian undertones appealed to important scientists and science policy makers at a time when they sought to shift the burden of expectation for industrial innovation from the private sector to academic research communities (2005, 179–186). Nanotechnology seems to have been first used regularly in a nonfuturist setting by chemist Richard Smalley at Rice University around 1992. A pioneer in the field of atomic clusters, the intermediate scale of matter between atoms and molecules, and the codiscover of C_{60}, the first buckminsterfullerene, Smalley employed the nano prefix as early as 1988 and began referring to *nanotechnology* in invited presentations around the summer of 1992 (Smalley 1990, 1998).

Drexler's influence in this respect and his relationship with Smalley in the early 1990s has been extensively chronicled (Regis 1995; Bensaude-Vincent 2004; McCray 2005; Berube 2006). Then a professed fan of the engineer, Smalley distributed copies of Drexler's works to Rice University's board of governors in support of his campaign to establish an institute for nanoscale science and technology. In a January 1993 letter to Michael M. Carroll and James L. Kinsey, the deans of Engineering and Natural Sciences, respectively, at Rice University, Smalley wrote that although Drexler's ideas were not especially useful for choosing research directions, the "central idea of nanotechnology"—the purposeful manipulation of matter on the nanometer scale—was or would shortly become a reality in a broad range of fields. He concluded that nanotechnology should be interpreted much more broadly, and that when this happened, interest among scientists would greatly increase (Smalley 1993).

By then, Smalley was abandoning the field of cluster spectroscopy in favor of carbon nanotubes, a form of fullerene first studied in depth by the NEC Corporation researcher Sumio Iijima around 1991. Smalley believed this material had strategic potential, especially when applied in existing energy storage and power source technologies (Smalley 2003). He devoted the remainder of his career to building institutions at Rice University and elsewhere dedicated to producing this substance. Hence, Smalley used the term *nanotechnology* both literally, to refer to nanometer science yielding nanoscale materials, and figuratively, as an "organizing principle" attractive to a broad new science constituency as he engaged in institution building (Mody 2010, 14).

The National Science Foundation (NSF) became the first U.S. federal science agency to explicitly use the *nano* prefix as a formal programmatic title when it launched the National Nanofabrication Users Network (NNUN) in 1994. This was an infrastructure pool for what was formerly known as microfabrication located at Cornell, Stanford, Pennsylvania State, and Howard universities and the University of California in Santa Barbara.

Several years later, a group of civil servants in the U.S. federal science policy apparatus began to use nanotechnology as a rubric for a discourse of revolutionary applied science. They adapted an expansive notion of the term embodying Drexler's technocratic assumptions and historical philosophy and Smalley's practical aspirations for special nanoscale materials. Their motive, suggested former NSF chief and Clinton science advisor Neal Lane, was to leverage expectations of practical applications of a range of novel materials to increase federal funding for the physical sciences, whose share of federal science resources had long been in decline relative to the health sciences (Lane and Kalil 2005, 50). Founded in November 1996 by NSF Engineering Directorate official Mihail C. Roco, the Interagency Working Group on Nanoscience, Engineering, and Technology (IWGN) enlisted Smalley, a co-recipient of the Nobel Prize in Chemistry for 1996 for the discovery of buckminsterfullerene, in their campaign in the late 1990s. In 1999, he promoted the idea of nanotechnology as a kind of general-purpose technology in congressional testimony.

These actors emulated Drexler in conceiving nanotechnology as less a discrete process or object than an idealized model of the social relations of innovation. Lane understood nanotechnology as expressing both current basic research (nano) and deep-future applications (technology) (Lane 2001, 96). Conceived thusly, nanotechnology was a form of science that bridged all the conventional divisions of the innovative process, one the interagency group promised would trigger an industrial revolution (IWGN 1999, vi).

In the 2000s, advocates of nanotechnology began to domesticate the term. The favored approach, as ever, was to associate with Feynman and his prestige and legitimacy as a Nobel laureate. President Clinton framed him as a founding father of nanotechnology in formally announcing the National Nanotechnology Initiative in January 2000 (Clinton 2000). Drexler, on the other hand, had long been a divisive figure in the science community. Increasingly criticized in a series of popular and professional journals of science in the early 2000s, he became a pariah.

Part of the reason for the backlash was that some proponents of nanotechnology feared that visions of its catastrophic potential sketched by Drexler and publicized by high-profile figures like Sun Microsystems cofounder Bill Joy might undermine the public's confidence in research conducted under this rubric (Berube 2006, 38–41, 49–80). To be sure, general awareness of nanotechnology remained low in the developed world in the 2000s (Satterfield et al. 2009, 752–758). And in fact, Smalley and Drexler shared similar metaphysical assumptions about nanotechnology, seeing biology and technology as mutually analogous. But they diverged on the question of agency. Drexler and his supporters adopted the literal metaphors of Cartesian reductionism, envisioning rigidly mechanical programmable assemblers as a means of enabling humans to completely control homogeneous, passive matter and reshape it into superior synthetic forms. In contrast, notions of dynamic nanotechnology developed by chemists and materials scientists like Smalley and George Whitesides were based on an active model of nature in which heterogeneous, recalcitrant matter could never be fully controlled but only coaxed into self-assembly (Bensaude-Vincent, 2004, 65-82).

Nano-prefixed terminology and the expression *nanotechnology* shared different trajectories as signifiers. The latter became part of the lexicon of science policy but not the lexicon of science practice; the opposite has generally occurred with the former. As an expression connoting a form of science with powerful technological implications, nanotechnology helped justify a contradictory brew of ideas relating to concerns of the economic values of science in the 1980s and 1990s, as McCray has suggested (2005, 192–195). Expressing the idea of a form of science with powerful technological implications, marrying the notions of radical physical and biological change with social stasis, nanotechnology served as an allegory for market forces, for it avoided the "centralized planning" of top-down fabrication (Sargent 2006, xvi). It is not clear what role nanotechnology will play as a signifier in the second decade of the twenty-first century. New actors may well imbue it with new meanings as they have done in the past. Alternatively, the word could pass into obsolescence and disuse, the fate of *nanophase*. Already there are signs that this is taking place, with *synthetic biology* emerging as a possible contender to replace *nanotechnology* in its role as an exemplar or symbol.

REFERENCES

Bensaude-Vincent, B. 2004. Two cultures of nanotechnology? *HYLE-International Journal for Philosophy of Chemistry* 10(2): 65–82.

Berube, D.M. 2006. *Nano-Hype: The Truth Behind the Nanotechnology Buzz*. Amherst, NY: Prometheus Books.

Choi, H., and C.C.M. Mody. 2009. The Long History of Molecular Electronics: Microelectronics Origins of Nanotechnology. *Social Studies of Science* 39(11): 11–50.

Clinton, W.J. 2000, January 21. "Address to CalTech on Science and Technology." Retrieved from http://pr.caltech.edu/events/presidential_speech/.

Drexler, K.E. 1986. *Engines of Creation:* Garden City, New York: Anchor Press/Doubleday.

Drexler, K.E. 1981. Molecular engineering: an approach to the development of general capabilities for molecular manipulation. *Proceedings of the National Academy of Sciences* 78(9): 5275–5278.

Gallo, J. 2009. The discursive and operational foundations of the National Nanotechnology Initiative. *Perspectives on Science* 17(2): 174–211.

Interagency Working Group on Nanoscience, Engineering, and Technology. 1999. Nanotechnology Research Directions; *IWGN Workshop Report*. Eds. M.C. Roco, R.S. Williams, and P. Alivisatos, Loyola College, MD: International Technology Research Institute, World Technology Division.

Johnson, A. 2009. Modeling molecules: computational nanotechnology as a knowledge community. *Perspectives on Science* 17(2): 144–173.

Krumhansl, J. A., and Y.-H. Pao. 1979. Microscience: an overview." *Physics Today* (November): 25–32.

Lane, N. 2001. The grand challenges of nanotechnology. *Journal of Nanoparticle Research* 3: 95–103.

Lane, N., and T. Kalil. 2005. The National Nanotechnology Initiative: present at the creation. *Issues in Science and Technology* (Summer): 49–55.

McCray, W.P. 2005. Will small be beautiful? Making policies for our nanotech future. *History and Technology* 21(2): 177–203.

Mody, C.C.M. 2011. Conferences and the emergence of nanoscience. Forthcoming chapter in *Social Life of Nanotechnologies*. Eds. B.H. Hawthorn and J. Mohr. New York: Routledge.

Mody, C.C.M. 2010. *Institutions as Stepping Stones: Rick Smalley and the Commercialization of Nanotubes*. Philadelphia: Chemical Heritage Foundation.

Mody, C.C.M. 2009. Introduction. *Perspectives on Science* 17(2): 111–122.

Mody, C.C.M. 2004. How probe microscopists became nanotechnologists. In *Discovering the Nanoscale*. Eds. D. Baird and J. Schummer. Amsterdam: IOS Press: 119–134.

Nordmann, A. 2009. Invisible origins of nanotechnology: Herbert Gleiter, materials science, and questions of prestige. *Perspectives on Science* 17(2): 123–143.

Regis, E. 1995. *Nano: The Emerging Science of Nanotechnology.* Boston: Back Bay Books.
Sargent, T. 2006. *The Dance of Molecules: How Nanotechnology Is Changing Our Lives.* New York: Thunder's Mouth Press.
Satterfield, T., M. Kandlikar, C.E.H. Beaudrie, J. Conti, and B. Harthorn, 2009. Anticipating the perceived risk of nanotechnologies. *Nature Nanotechnology* 4 (November): 752–758.
Schummer, J. 2006. Gestalt switch in molecular image perception: the aesthetic origin of molecular nanotechnology in supramolecular chemistry. *Foundations of Chemistry* 8: 53–72.
Smalley, R.E. 2003. "Online Newshour: The Future of Fuel: Advances in Hydrogen Fuel Cell Technology." Interview by T. Bearden. October 20. Retrieved from www.pbs.org/newshour/science/hydrogen/smalley.html.
Smalley, R.E. 1993. Letter to Carroll and Kinsey, January 21, Series II, Correspondence, 1992–1996, Box 3, Folder 3. Richard Smalley Papers, Woodson Research Center, Fondren Library, Rice University, Houston, TX.
Smalley, R.E. 1990–1998. "Presentations: Listings," Series IV: Personal, Box 4, Folder 8, Richard Smalley Papers, Woodson Research Center, Fondren Library, Rice University, Houston, TX.
Subcommittee on Nanoscale Science, Engineering, and Technology, Committee on Technology, National Science and Technology Council. 2010. The National Nanotechnology Initiative: Research and Development Leading to a Revolution in Technology and Industry: Supplement to the President's FY 2011 Budget.
Taniguchi, N. 1974. On the basic concept of "nanotechnology." *Proceedings of the International Conference on Production Engineering, Tokyo 1974 (Part II).* Tokyo: Japan Society of Precision Engineering, 18–23.
Toumey, C. 2005. Apostolic succession: does nanotechnology descend from Richard Feynman's 1959 talk? *Engineering and Science* 1/2: 16–23.

Acknowledgments

This material is based upon work supported by the National Science Foundation under Grant No. SES 0531184 and 0938099. Any opinions, findings, and conclusions or recommendations expressed in this material are those of the author and do not necessarily reflect the views of the National Science Foundation.

References

Beck, U. 1992. *Risk Society: Towards a New Modernity*. Trans. M. Ritter. London: Sage.

"Nanotechnology consumer products inventory." 2010. Retrieved from www.nanotechproject.org/inventories/consumer.

"What is nanotechnology?" n.d. Retrieved from www.nano.gov/html/facts/whatIsNano.html

2

Introducing Nanotechnology to the Public

Nanotechnology is a class or group of technologies enabled by advances in microscopy and related developments in engineering that allow us to visualize and manipulate matter at this extremely small scale. It is not a particular technology but a group of technologies enabled by a particular set of technological (engineering) capacities and distinguished by their exploitation of the properties of certain materials "at the nanoscale," which can be different from their properties in their naturally occurring forms. For example, very small particles have a larger surface-to-mass ratio, and therefore can behave differently in chemical reactions. At one time, it was common to refer to "nanoscience and nanotechnology" in order to suggest a distinction between these parallel to the general distinction between science, which is theoretical knowledge and the research associated with increasing that knowledge, and technology, which is something created, built, or manufactured—the development of which is often dependent on our scientific knowledge—for a human purpose. It is now common to use the term *nanotechnology* to refer to both of these, given that both the science and the technology associated with the nanoscale are evolving closely together.

Some argue that the particles that create air pollution are also, in terms of their size, nanoparticles, and biologically important molecules such as DNA might also be seen as nanoparticles. However, when we talk about nanotechnology (or any kind of technology, for that matter), we are generally referring to material that has been modified or otherwise deliberately manipulated by humans for certain purposes. While particulate matter in the air is a result of human processes (such as the operation of internal combustion engines or the burning of coal), the goal is not to create the particles, which are an unfortunate by-product rather than something deliberately manufactured to meet our needs. DNA molecules can also be manipulated and modified through genetic engineering processes, but even genetic engineering does not generally create DNA. On the other hand, the DNA made through synthetic biology, which involves the artificial generation of particular types (sequences) of DNA using DNA synthesizer technology, does come very close to the usual definition of *nanotechnology*.

As a broad class of technology and not a single process or item, nanotechnology can be unusually difficult to grasp conceptually, especially for nonspecialists. While this is also true of biotechnology, nanotechnology seems especially difficult to imagine because of the many different forms it can take, as well as its ultimate invisibility to the human eye. Scholars have sometimes argued that the invention of the term *nanotechnology* actually represents a rather artificial categorization, one arguably first made for the political purpose of gaining support for major investments of research dollars but not necessarily a category that scientists and engineers would otherwise have found uniquely useful or natural as a class or category of technology or science. Yet various, although widely diverse, contemporary engineering advances depend on the same emerging set of visualization and manipulation capacities "at the nanoscale," so there is a certain logic to grouping these together. This grouping can, however, be a problem from a risk communication point of view, because some kinds of nanotechnology and nanoparticles pose very low levels of risk (or at least low levels based on current knowledge), while evidence is accumulating that others can be problematic.

Imagining the Nanoscale

In some ways, envisioning nanotechnology requires considerable imagination. The human eye cannot see materials at the nanoscale. As a result, some degree of controversy has arisen over how nanomaterials or other forms of nanotechnology should be represented (e.g., in illustrations), because these cannot ever actually be seen except through the use of advanced microscopy technology. How do we imagine what something would look like if we could see it, when in fact we cannot (and can never) see it with our eyes, but can only see an artificial representation of it constructed by a machine? Many magnified illustrations add color or exaggerate (that is, magnify, sharpen, or otherwise clarify) these molecular patterns and shapes in attempts to make the properties of nanostructures more obvious. But what, ultimately, is the essential visual nature of these structures and materials? Creative illustrations designed to represent nanotechnology may be viewed by some as distortions that could have an effect on perceptions of nanotechnology's nature and value, including its usefulness, its harmfulness, its attractiveness, or all of these and more. What something that cannot actually be seen would look like if it could, after all, become directly visible is a conundrum that challenges our idea of an observable reality.

It now appears clear that some forms of nanotechnology may pose risks to human health and environment, while others behave entirely differently and may pose little or no risk. For example, carbon nanotubes—long nanoscale tubes made of carbon atoms—have been implicated in the development of mesothelioma, a rare form of cancer also associated with asbestos (Poland et al. 2008; Takagi et al. 2008). Yet as of now, even experts do not always know how to tell ahead of time the difference between a type of nanomaterial that is harmful and one that is not. Right now a great deal of research effort is being invested in figuring out how to identify the potentially harmful products of nanotechnology and distinguish them from the products that are not harmful. Every form of nanotechnology is unique. To compound the confusion, the word *nano* or *nanotechnology* is sometimes used to advertise or market products that do not necessarily have nanoscale components or dimensions. A European cleaning product called "Magic Nano," for example, seemed to cause health problems in some users, but on closer inspection it is not clear that actual nanoparticles (by the broadly accepted definitions discussed above) were even involved; the manufacturer may have just tagged on the "nano" label as a selling point (von Bubnoff 2006). This has obvious potential for adding to public confusion.

Observers have sometimes worried that in the wake of such cases (whether the nanotechnology connection is technically real or not), some people will tend to lump all nanotechnology together, believing that if one form is risky or otherwise objectionable then all others must be suspect as well. There is no clear evidence that this has happened so far. However, having witnessed the emergence of popular concern over nuclear power, stem cell research, genetically modified foods, agricultural chemicals, and a host of other new technologies over many decades, some scientists, engineers, and policy makers have come to believe that ordinary people are unreasonably risk averse and will react to any technology with fear and resistance. This is also not necessarily the case. In addition, of course, some public fears ultimately do turn out to be well justified, despite protestations to the contrary of both industry and government. Further, popular judgments about the acceptability of a new technology are often based as much on ethics and values considerations as on the simple misperception of risks, narrowly defined. The complex challenges involved with communicating risks to broader publics in effective and yet honest and transparent ways have certainly been highlighted by experiences with the introduction of other technologies that have proven to be unexpectedly controversial, and occasionally dangerous. Before Rachel Carson's *Silent Spring* (1987, first published in 1962), environmental harm associated with DDT and other pesticides went unrecognized. Nuclear power, long officially proclaimed as safe, brought us the events at Three Mile Island and Chernobyl, reinforcing public suspicion about official pronouncements of the technology's safety. In the United Kingdom, government scientists initially insisted that "mad cow disease" did not affect humans. It is little wonder that some people have become cautious about technology as a result of this history.

Perhaps it is natural for many in the scientific and engineering communities, and very often the science policy community as well, to assume that if the general population of people had possession of all the scientific facts—if their levels of scientific knowledge and science literacy were improved, in other words—they would not raise so many obstacles to the acceptance of new technologies. Critics see this view (referred to as the *deficit model* of public understanding of science) (Ziman 1991; Miller 2001) as based on an erroneous view of opinion formation that is largely in contradiction to available evidence, the mistaken idea that a better grasp of the relevant scientific facts would make most people think like scientists and engineers, as well as dissolve most of what the scientific community may see as unfounded objections to technological advancements.

The reality of the situation is generally much more complex and nuanced. After all, the value systems of scientific experts can be expected to lean toward embracing science and technology. Although it is inherently good for people in economically and technologically developed societies to understand as much as possible about the science and technology that is vital to their way of life, it is just not true that all objections to technology result from a lack of scientific understanding. Most such highly developed societies, including the United States, much of Europe, and (increasingly) parts of Asia, are very pluralistic in terms of people's ethnic backgrounds, religious and cultural beliefs, and related value systems. Globally, migration is increasing. Further, some values and beliefs, including contradictory ones that may coexist within the same society, may support some aspects of technological development but question others, greatly complicating the question of popular receptivity to particular technologies.

The bottom line here is that not all objections to emerging technologies stem from scientific illiteracy. As larger segments of the scientific, engineering, and policy communities recognize this, more work is being done to understand and, ideally, to respond to people's questions and concerns before emerging risks and controversies over those risks become unmanageable. We have some "breathing room" in the case of nanotechnology, as public acceptance appears, so far, to be relatively assured.

Another way of seeing the "big picture" of public receptivity to new technology is to examine more closely the adoption patterns for technologies that already exist. It is immediately clear that contemporary society enthusiastically embraces some technologies (e.g., cell phones) despite uncertain and still unknown risks, and without fully understanding just how they work. Even though cell phone use while driving is blamed for a huge number of car accidents, and even though the effects on biological cells of exposure to the electromagnetic fields generated by cell phones and other similar electronic devices are not yet entirely understood (Redelmeier and Tibshirani 1997), we tend to love our cell phones anyway. It is abundantly obvious that they are useful to us and that they make our lives more convenient and more enjoyable. In our increasingly mobile society, they allow us to manage our very busy lives and keep in touch with our families, responding to our deepest social needs and our most important social values. The fact that most of us could not explain just exactly how a cell phone works, and that the use of these phones is not known with absolute certainty to be entirely without risk, seems largely irrelevant, given their obvious usefulness and value in people's everyday lives. In other words, people demonstrate every day that they are willing to assume small risks, even uncertain ones, if the technology in question has important benefits for them.

Similarly, public resistance to genetically modified foods probably resulted less from poor understanding of the underlying principles of genetics and other science having resulted in exaggerated perceptions of risk as it did from other factors, including concern about long-term environmental effects (which remain imperfectly understood) and about economic and other societal impacts (such as the effects on small-scale family farming, which also remain imperfectly understood). Even small challenges to cultural values can result in significant resistance to new technology in the absence of clear evidence of tangible benefits. And where there is no obvious consumer benefit to outweigh even limited risks and uncertainties, as in the case of genetically modified foods, concern over even minor risks can prevail (Lang and Hallman 2005). In most cases, people have not seen their food become more nutritious, cheaper, safer, or tastier as a result of genetic modification, which has evolved in directions more to the immediate advantage of agricultural and agribusiness concerns than that of food consumers. Thus, even a very small increase in perceived risk to health or the environment was not likely to be overlooked by those consumers.

Further, some people may reject genetically modified foods because of their belief that "natural" foods and "natural" farming are inherently more valuable, a perspective that is clearly as much a reflection of social values as a reflection of scientific knowledge (or, conversely, of any lack of that knowledge). Some might argue that this view reflects an idealistic, even romantic, notion of the nature of modern large-scale food production, as much influenced by the industrial revolution as any other mass manufacturing process has been. Nevertheless, people tend to cling to their preferences. In addition, on the international level, Europeans have been more resistant to genetically modified foods than were people in the United States, in part because these foods were initially being introduced from elsewhere—specifically by foreign (i.e., U.S.) interests. Any group's sense that technological change is forced on them without discussion or assent seems a sure recipe for popular resistance, and these foods were introduced without any broad public discussion, so that even in the United States (biotechnology's homeland in many ways), some people were shocked to find out that the nature of some foods they were already consuming had been changed without their knowledge.

Emerging Public Perception

Nanotechnology does not need to follow the path of biotechnology in terms of public perception. The *deficit model* idea that public resistance to technology is always and necessarily the result of poor understanding, prejudice, or exaggerated perceptions of risk generally implies that science education will ultimately eliminate most resistance to technological change, a view that seems to reemerge with every emerging technology. Yet opponents of a particular technology, whether genetically modified food crops or nuclear power, often know as much or more about that technology as do those fully willing accept it. Somewhat ironically, given the growing evidence that some nanotechnology involves risks, popular perception may in this case understate, rather than overstate, these risks—dynamics we do not fully understand but that again point to the role of cultural beliefs and social values in risk perception.

In fact, sound education about science and technology often teaches about risk as well as benefit and carries no certain guarantee of universal acceptance, in the end clearly a function of social values as well as knowledge of scientific facts. Today we understand technology as a product of social choices, not something that develops on its own. A small proportion of differences in popular attitudes toward controversial technologies can be explained in terms of possession of scientific knowledge, but most such differences ultimately have other causes (Sturgis and Allum 2004). In a modern democracy, we hope that as many people as possible are educated enough about matters of public policy to engage in constructive discussion, form reasonable opinions, and propose wise decisions, or at least this is the ideal. Collective wisdom results from open and collective discussion, assuming open access to information (sometimes, for politics, referred to as the *marketplace of ideas* model).

American educational philosopher John Dewey (see 1927) has been among the best-known proponents of the idea that "ordinary" citizens, given appropriate skills, could reach rational and wise opinions about even complex policy questions. This view applies to today's discussions of public engagement, opinion formation, and collective decision making about science and technology, as well as to other branches of participatory democracy. Of course, scientific and technical expertise on particular topics will remain concentrated in a small percentage of the population; as a practical matter, we are not all going to become scientific experts, and even scientific experts are specialized, meaning that their expertise is generally quite narrow and specific. The rest of us, a majority of whom are not scientists and who are certainly not scientific or technical experts on every science-related topic, are nevertheless likely to argue for our right to weigh in. We will choose to buy products or not, at a minimum, and we may also develop strong opinions about society's investment in particular technologies, opinions we can express not just in our voting but also in advocacy for or against certain policies (and, ultimately, the adoption or rejection of particular technological developments).

However, the ethical propriety of consulting broader publics specifically for the strategic purpose of heading off criticism is prima facie questionable; furthermore, there is no guarantee this strategy will work, and many historical examples exist where it has not. Conversely, not consulting with broader publics is a likely path to controversy. It is in the interests of the scientific, engineering, and policy communities to take public opinion into account in making technology policy, and it is the ethically appropriate thing to do. Democratic theory generally recognizes that all citizens have the right and the obligation to become engaged in policy making in various ways and to varying degrees (even though not all of us can become engaged in all issues), and this book reflects this underlying assumption about the appropriate relationship between technology and society. Whether or not one accepts this assumption (a value-based perspective), the fact remains that ignoring public opinion invites popular resistance to the adoption of technological change.

Of course, opinions may differ on just how much the public should be involved in policy decisions about science and technology. Broader publics can also be wrong: Climate change is in fact taking place, and vaccination in fact prevents certain important diseases. How to accommodate views to the contrary is a largely unsolved issue of political philosophy and, occasionally, of law (e.g., when parents refuse to send their children to school or have them treated for disease because of their own belief system). But even for those who would like to see most decisions about science and technology made primarily by experts, the fact is that the broader publics of which modern society is composed will often insist on having a say, nevertheless. "Upstream" public engagement in the development and deployment of emerging technologies is now recognized as both a public good and a practical necessity. The interests of the promoters of new technology and the promoters of good democratic practice have largely converged on this point.

The invisible and (for many) therefore abstract and intangible nature of nanotechnology, and the use of this term to describe such a wide variety of processes and applications, has made it difficult to communicate about this subject to nonspecialists, and many people remain largely unaware of either nanotechnology's promise or its perils. Some observers feared that early fictional portrayals such as Michael Crichton's novel *Prey* (2002), which depicted robotic nano-insects gone wild, or concerns about tiny self-replicating nanomachines turning the universe into "grey goo" (voiced by the United Kingdom's Prince Charles among others) (Radford, 2003) would terrify a public ignorant of nanoscience. But this did not actually happen, or at least has not happened yet. (Oddly, readers of *Prey* are on average more pro-nanotechnology than not, if only because science fiction fans are more likely to be technophiles.) (See Cobb and Macoubrie 2004.)

Instead, as evidence is building that some nanotechnology, under some circumstances, might possibly prove harmful to humans, animals, or ecosystems, and as it becomes increasingly apparent that existing regulatory systems may not be responding very quickly to these possibilities, a new picture emerges. Instead of being unreasonably frightened, people seem remarkably complacent. Unlike biotechnology, nanotechnology does not seem to induce an initially fearful reaction from most. Perhaps this is because, as a participant in one U.S. focus group on the subject of nanotechnology put it, "materials don't have ethics" (Priest and Kramer 2008); that is, the artificial manipulation of nonbiological materials does not seem to induce the same sort of initial public reaction as the artificial manipulation of biological materials such as DNA. While people do recognize some ethical issues surrounding nanotechnology, such as how both the benefits and the risks will be distributed, the same cultural foundation that seems to have produced resistance to some forms of biotechnology seems largely to have accepted nanotechnology.

Social scientists who have studied public reactions to risks have coined the phrase *social amplification of risk* to describe the tendency for some risks to become exaggerated as they become the subject of news stories and the object of advocacy group concerns (Kasperson et al. 2003). According to this idea, various social institutions such as the news media serve as *amplification stations* that, as they send risk information out into society, tend to augment or amplify the power of these messages. Much of the attention to societal reactions of risk, however, has focused on cases of exaggeration rather than (as with nanotechnology or, say, climate change) cases in which documented risks are widely ignored. Failure to recognize and take action about risks is just as problematic, or arguably even more so, as the tendency to exaggerate them, but has less often been studied. Further, the successful deployment of the socially beneficial applications of nanotechnology is ultimately dependent on appropriate societal reactions with respect to evaluation, regulation, and management of those risks. Some observers continue to worry about what will happen as information about risks becomes more broadly disseminated to various publics.

The research on social amplification has also considered cases of social attenuation dynamics through which some risks tend to be downplayed or ignored, even while other amplified risks receive ample, sometimes disproportionately excessive, attention. Yet the social and behavioral theory that might fully explain why one risk might get a lot of public attention while others are ignored has not been well developed. It is reasonably obvious that we tend to ignore or downplay the risks of technologies that provide us with clear benefits (such as cell phones or automobiles).

But nanotechnology's most substantial benefits lie mostly in the future. Although a handful of advocacy groups have expressed concern about nanotechnology's environmental and health effects, and in the United States and elsewhere increasing levels of government research dollars are being invested in investigating those effects, many people remain unaware of both the risks and the benefits of this "sleeping giant" technology. This is where risk communication comes into play. In cases of amplified risks, risk communicators are often asked to develop strategies to reduce what some see as exaggerated public concerns, but in cases of attenuated risks, their role is likely to be somewhat different.

Risk Communication for Twenty-First Century Democracies

Alongside tool use, communication is often cited as a defining characteristic of what it means to be human. Although other species of primates (notably chimpanzees, our closest relatives) use simple tools and can be taught to communicate via symbols, this is rare elsewhere in the animal kingdom, and as far as we can discern, no other species carries these capacities as far as human beings do. In addition, both the control of tools and the control of communication processes are critical sources of power and influence in all human societies. Communication technology in particular, embodying the intersection of these two sources of power, has played a crucial role in the organization of complex societies, even empires, ranging from the emergence of early civilizations (see Innis 2007) to the invention of the movable type press in Europe (see Eisenstein 1994), and from the extension of the British, French, German, and other European empires to the wide influence of today's communication satellites (see, e.g., Schiller 1976).

Today, global media power is increasingly concentrated in a handful of media-oriented corporations. Extensive discussion has taken place among media scholars about the way governmental and corporate control over communication technologies is maintained and distributed, which varies among nations. In the United Kingdom, the British Broadcasting System has historically had a very close relationship with the U.K. government. In the United States, formal legal and political control of communication media through government is relatively limited compared to policies in other countries, arguably to the advantage of corporate interests. Yet even in the U.S. system, there are laws designed to assure the availability of multiple points of view. To highlight just one crucial example, joint operating agreements legally enforced by U.S. courts have guaranteed that more than one newspaper will continue to operate in some major markets, even though specific elements (such as their respective advertising departments) may have been merged to create economic efficiencies.

The underlying principle here is the importance, to a successful democracy, of a multitude of editorial voices informing us on issues of the day. Some would argue that this policy (joint operating agreements) is too little, too late, given the extent of corporate dominance of most modern media channels around the globe. But the ongoing enforcement of such policies helps illustrate that the necessity of providing multiple perspectives continues to be recognized by those concerned with the relationship between democratic practice and the availability of a variety of points of view. Many observers have hoped that the proliferation first of cable and satellite TV, then of new forms of Internet-based communications, will guarantee the perpetuation of choice from a range of alternatives. So far, the remarkable sameness of most media material has not entirely fulfilled these hopes, but the international spread of the Internet and other forms of electronic media have at least made it much more difficult to suppress the expression of political dissent.

As this chapter should illustrate, risk is most often a matter of collective perception and expectation, as well as scientific fact. In other words, risk—or, at a minimum, risk perception—is a matter of opinion. As is true for political news, access to a variety of opinions and judgments about risk issues helps democracy flourish. This is not to say that the scientific facts (including the results of expert risk analysis) can be ignored, or that these are not important. However, risk specialists now recognize the value-laden character of the judgments that must be made about the trade-offs involved in accepting or rejecting particular risks, given that many risks cannot be accurately estimated with a known level of certainly. Even where accurate formal estimates can be made, reasonable and well-informed individuals may still disagree about whether a given level of risk is acceptable or unacceptable.

In line with the dynamics that have been described in the preceding sections, we can define some of the characteristics of good risk communication for the type of twenty-first century democratic society (characterized by high levels of economic and scientific development) into which nanotechnology is being introduced. A key characteristic involves recognition of the two-way nature of public communication efforts, meaning that the perceptions and preferences of a broad array of publics need, ideally, to be accommodated by decision makers (Grunig and Hunt 1984). In communication studies more generally, we have moved from an era in which communication was defined by a linear, one-way paradigm in which it was seen as a process of transferring information from one person (or from a single point, or from one group) to another, to an era in which various other ways of thinking about communication have emerged and begun to dominate. Communication is no longer thought of as linear but as dynamic, as two-way or multidirectional, and as influenced by culture, social structure, and other elements of the interpretive context.

Just as communication is generally best thought of as complex and dynamic rather than linear, communication about technology, including communication about nanotechnology, should also be thought of as complex and dynamic. In other words, it is no longer understood as being just about translation anymore, as the old linear model conceptualized it, but about the importance of creating the conditions for constructive social dialogue. Disseminating information and hoping audiences will take it up is no longer good enough. In the era of the Internet, no one has complete control over the flow of information, and many other factors influence the formation of public opinion, whether about technology or about any other decision confronting society.

Much has been written about the ways that new technologies are diffused and adopted (see, in particular, Rogers 2003). Scholars have pointed out that some individuals (those Rogers called "early adopters") are generally predisposed to adopt any new technologies that come along, whatever type of technology we are talking about. We all know someone who always seems to be the first to want to own every new gadget. Others are slower to embrace them, wanting to wait until the functionality, and price, of new technologies become more stable. (These people are sometimes called "late adopters" or even "laggards," in Rogers' terminology.) This process has political elements, not only psychological ones, with different interests and perspectives within society serving to accelerate or slow technology's widespread adoption based on a variety of perceived goals and interests.

In this context, competing political philosophies have evolved about how this process ought ideally to play out, with some observers urging precaution until the risks to humans, other species, and the ecosystem are well understood. Others, believing that new technology of all kinds is inherently valuable and necessary to our well-being, economic vitality, and general quality of life, believe we should be careful to avoid policies that might slow the progress of adoption. While some believe early and active regulation is necessary to assure consumer safety and protect the innovation process, others believe that marketplace forces should be allowed to prevail without interference. While some associate technology with their suspicions about the power of big business and multinational corporations, others think first of the much smaller-scale innovation that takes place on the local level. These perspectives arise in discussion of communication technology, of nanotechnology, and of all other emerging technologies.

In practice, these differences in perspective clearly depend as much on one's more general political and economic assumptions as on any actual characteristics of a particular technology. (At the same time, every technology seems unique, in terms of what interest groups come into play and the ways in which public opinion is actually formed.) One result is that risk communication often takes place in a political context that has become highly charged—one might almost describe it as a minefield. Further, some technologies (such as human cloning, for example, or, for some, embryonic stem cell research) can challenge some of our fundamental ethical values as well. Given all these competing ideas and more, what is the best way to talk about the risks of technology, and what are the goals of doing so? Why do we even have a field called "risk communication" to begin with, when technologies are what bring us so many benefits as well?

Of course, professional communicators who work in the areas of science and technology are also interested—very often *most* interested—in conveying the benefits, as well as the risks, of new developments. Often these individuals are in jobs within institutions such as universities, research institutes, and corporations large and small that are in the business of promoting the adoption of new technologies and practices. It is no contradiction to suggest that popular reactions should be of concern here; in the end, whether justified or not, popular evaluations determine which technologies will become controversial, and which risks are worth the associated benefits. Other risk communication specialists may work for consumer groups or other advocacy groups that want to make sure that risks are not overlooked in the process of discussing and promoting benefits. It is around the risks of technology (real or perceived), not only its benefits, that the political, regulatory, and social controversies surrounding new technology tend to emerge, whether the subject is a specific consumer product or a broader regulatory initiative. Nanotechnology, like almost all new technologies, does carry risks, and the novel properties that infuse so much hope into the development of nanotechnology's various applications also create an unusually novel set of risks that we are still learning how to identify and manage (see, e.g., Shatkin 2008).

Of course, it is not just risk communicators who participate in shaping public opinion about technology and its risks and benefits. Journalists who write about science and technology also play a strong role. Even though working in a field that prides itself on maintaining objectivity about social and political issues, science journalists often enter the field out of a belief that developments in science and technology have inherent value to society. But like all good journalists, they are concerned with exposing risks and other negative elements, as well as describing benefits. Their commitment to objective reporting shares common ground with society's commitment to scientific objectivity (Nelkin 1987), but this does not mean their work is always entirely value free.

Theories of democratic pluralism assert that this broad mix of promotional, advocacy, and journalistic efforts will ultimately promote the application of wisdom and good decision making. This is what is meant by the *marketplace of ideas* concept, a conception of the public sphere as composing an open marketplace in which different ideas compete on a level playing field for acceptance or rejection by citizens of a democracy. There are limits to this idea, of course. Power and resources are not equally distributed, and the playing field is therefore simply not level to begin with. Yet this is the environment in which risk communication takes place. Philosophically, we might hope for an ideal something like that articulated by Habermas (1985) in which all parties involved communicate on a relatively equal footing. The day-to-day reality is that this ideal remains more than a little elusive, however.

PERSPECTIVE: PROFESSIONAL PRINCIPLES OF RISK COMMUNICATION

Joye Gordon

"What's that?" is the likely response if you tell someone you are in the business of risk communication. In a broad sense, risk communication is communication about physical hazards. But that can entail anyone from lead abatement officers to lobbyists and journalists. There is little wonder that confusion surrounds risk communication given the debate over the understanding of risk.

From a physical hazards perspective, risk is a function of the likelihood and magnitude of negative consequences. However, such calculations to predict outcomes are value driven, inherently uncertain, and invite debate. Despite a desire to rely on empiricism to maximize safety, societies ultimately must make decisions about hazards under conditions of uncertainty. Should society ban or encourage some emerging technology? Where will the landfill be located? Should laws mandate seatbelts and helmets? And must one really submit to body scanners before boarding an airplane? How such questions are answered is a function of debate and communication more than a function of calculations of risk assessors.

In 1989, the National Research Council formalized a definition of risk communication as the exchange of information and opinions involving multiple messages, not strictly about risks, relating to risk management. The definition emphasizes that risk communication is a two-way and interactive exchange. Communication efforts that focus on maximizing message impact or communication efforts that exist to gain compliance do not meet the normative intention of the

National Research Council's definition that acknowledges the right of and the need for citizenry to actively participate in debates. As such, risk communication can be understood as all messages and interactions that bear on risk decisions, including statements of values and opinions.

The breadth of this definition is reflected in both the professional world and academia. Professional communication specialists may be asked to develop strategies to motivate risk reduction activities or to facilitate public hearings. Position announcements may ask for skills from social networking to the ability to navigate specific federal structures such as the National Flood Insurance Program. Academic fields addressing risk communication run the gamut from crisis communication to disaster sociology to emergency management and industrial hygiene. The CAUSE model developed by Katherine Rowan provides a mechanism for putting the various challenges and activities of professional risk communicators into perspective (see Rowan et al. 2003, 2009).

The CAUSE model and mnemonic identify the inherent tensions associated with risk communication, allowing for an examination of dilemmas professional communicators must address if risk communication is to successfully facilitate pubic debate. The C represents confidence; the A, awareness; the U, understanding; the S satisfaction; and, finally E for enactment. Ultimately, if satisfactory decisions evolve from the risk communication process, stakeholders much achieve some level of confidence and trust, comprehension, and agreement before long-term, sustainable stasis of behavior or regulation can exist.

Since the evolvement of the concept of ethos, communicators have known that successful communication depends upon the reasonable expectation of trust. Despite the attention that the trust, or speaker credibility, variable has received from communication studies, confidence remains elusive in many communication situations, especially situations involving hazards where risk managers and those affected hold differing perspectives. For example, despite ample empirical evidence supporting effectiveness, many resist vaccinations, fearing side effects, claiming ineffectiveness, opposing pharmaceutical companies, questioning morality, or doubting governmental intentions. Since the eighteenth century when vaccination began, these debates have impacted individual willingness and regulatory outcomes. When widespread inoculation is needed to prevent disease, the tension between individual liberties and public good is tested. If populations do not trust medical providers, vaccine suppliers, or governmental recommendations, the technology that saves lives goes underutilized, and preventable hazards continue to adversely affect populations. As such, building confidence remains a challenge for professional communicators even in familiar situations where statistical uncertainty is relatively low. As a necessary prerequisite to successful communication, trust and confidence are the first component of the CAUSE model.

The second and third components of the model are awareness and understanding. Creating awareness of hazards and deepening understanding are required to facilitate informed participation. Researchers have focused much attention on the best ways to reach audiences and to compose messages to maximize comprehension. Professional risk communicators are challenged to facilitate meaningful comprehension of sometimes complex and confusing information.

Fourth in the CAUSE model is the component of gaining satisfaction. In simplest terms, satisfaction can be equated to agreement. It is a state in which participants can feel mentally compatible with outcomes. In free societies, agreement, not force, precedes both governmental action and individual behavioral outcomes.

Finally, the CAUSE model concludes with enactment, the voluntary adoption of behavior or regulation concerning some hazard. People choose to receive vaccinations, or demand insurance coverage for vaccinations, or require vaccination for entry into public schools, or discontinue vaccinations. Even failure to respond is an enactment of neglect, of allowing a hazard to go unaddressed, often allowing hazards to escalate. Ultimately, risk communication should result in decisions and responses addressing some hazard in a manner acceptable to society.

In sum, the CAUSE model provides a pragmatic list of the components required for meaningful, sustainable, acceptable responses to hazards in society. Confidence, awareness, understanding, satisfaction, and enactment ideally are achieved through a risk communication process respecting participants' contribution to public debate. However, ideal processes and the actual practice of risk communication are often at odds.

Professional risk communicators are often motivated and approach hazard situations with a "solution" preceding the definition of the "problem." Economic motivations systemically provide more resources toward communication efforts aimed at proposal acceptance as opposed to public debate. As a society that values science, the technocratic approach may be hailed above human values as the proper way to approach our world. Public interest is often defined as valuing safety and security over personal liberties. The pejorative definition of risk communication as the practice of gaining compliance is not without merit. Case study after case study points to situations where public interests were not served and science was used as a tool to justify enacted responses when public acceptance was not achieved. Despite this dismal portrayal, risk communication is not inherently unethical. Like all communication endeavors, it requires careful attention to ideals of free flow of information and vigorous participation of stakeholders.

Definitions such as the one proposed by the National Research Council, along with codes of ethics offered by professional organizations contribute to an environment where risk communication is advanced as a contributor to the democratic process. But perhaps most influential on the practice of risk communication has been a host of legal and procedural mandates that prescribe communication activities. So prevalent are such rules that professional risk communicators spend a good amount of time and effort ensuring compliance with such mandates.

Most notable are acts and laws that ensure transparency, participation, and accountability. The Freedom of Information Act, the Right to Know Act, the Toxic Release Inventory, and the Clean Water Act are examples of protocols requiring transparency and mandating that various industries and governmental institutions make information available to citizens. The National Environmental Policy Act (NEPA) as a national policy requires all federal governmental agencies to prepare Environmental Assessments and Environmental Impact Statements relating to proposed actions. The process of preparing these reports requires opportunities for public comment. Although the intent of NEPA was environmental enhancement, the act institutionalized protocol that ensures citizens have opportunities to participate in debates about federal government actions. Finally, a host of rules and precedents demand accountability from those introducing hazards. Citizens have a Right of Standing in courts and the right to sue for recourse of damages suffered. Concerns that marginalized populations suffer more than benefit from risky actions led to the 1994 Executive Order on Environmental Justice. For professional risk communicators, meeting procedural guidelines for communication efforts requires intimate knowledge with a host of regulatory and legal requirements concerning the acts of communication.

Risk communicators face an ethical responsibility of facilitating democratic processes through free and open flow of communication.

Because hazard situations are inherently associated with tensions and dilemmas, the CAUSE model provides a pragmatic mechanism to identify and address the components of the communication process required to facilitate acceptable responses. Finally, professional risk communicators must be able to navigate federal structures and manage communication in accordance with the rules and regulations governing communication activities. From lead abatement officers, to lobbyists, to journalists, communicators that concern themselves with communicating about physical hazards are in a unique position, bridging the physical and social worlds and facilitating how we as a society manage hazards and threats.

REFERENCES

Rowan, K.E., C.H. Botan, G.L. Kreps, S. Samoilenko, and K. Farnsworth. 2009. Risk communication education for local emergency managers: using the CAUSE Model for research, education, and outreach. In *Handbook of Crisis and Risk Communication*. Eds. R.L. Heath and H.D. O'Hair. New York: Taylor & Francis, 168–191.

Rowan, K.E., L. Sparks, L. Pecchioni, and M. Villagran. 2003. The "CAUSE" model: A research-supported guide for physicians communicating cancer risk. *Health Communication: Special Issue on Cancer Communication* 15: 239–252.

References

Carson, R. 1987. *Silent Spring*. Boston: Houghton Mifflin.

Cobb, M.D., and J. Macoubrie. 2004. Public perceptions about nanotechnology: risks, benefits and trust. *Journal of Nanoparticle Research* 6(4): 395–405.

Crichton, M. 2002. *Prey*. New York: Harper Collins.

Dewey, J. 1927. *The Public and Its Problems*. New York: H. Holt.

Eisenstein, E.L. 1994. *The Printing Press as an Agent of Change: Communications and Cultural Transformations in Early-Modern Europe*. Cambridge: Cambridge University Press.

Grunig, J.E., and T. Hunt. 1984. *Managing Public Relations*. New York: Holt, Rinehart and Winston.

Habermas, J. 1985. *The Theory of Communicative Action: Reason and the Rationalization of Society*. Trans. T. McCarthy. Boston, MA: Beacon Press.

Innis, H.A. 2007. *Empire and Communications*. Lanham, MD: Rowman and Littlefield.

Kasperson J.X., R.E. Kasperson, N. Pidgeon, and P. Slovic. 2003. The social amplification of risk: assessing fifteen years of research and theory. In *The Social Amplification of Risk*. Eds. N. Pidgeon, R. E. Kasperson, and P. Slovic. Cambridge: Cambridge University Press, 13–46.

Lang, J.T., and W.K. Hallman. 2005. Who does the public trust? The case of genetically modified food in the United States. *Risk Analysis* 25(5): 1241–1252.
Miller, S. 2001. Public understanding of science at the crossroads. *Public Understanding of Science* 10(1): 15.
Nelkin, D. 1987. *Selling Science: How the Press Covers Science and Technology*. New York: W.H. Freeman.
Poland, C.A., R. Duffin, I. Kinloch, A. Maynard, W.A.H. Wallace, A. Seaton, V. Stone, S. Brown, W. MacNee, and K. Donaldson. 2008. Carbon nanotubes introduced into the abdominal cavity of mice show asbestos-like pathogenicity in a pilot study. *Nature Nanotechnology* 3(7): 423–428.
Priest, S., and V. Kramer. 2008. Making sense of emerging nanotechnologies: how ordinary people form impressions of new technology. Paper presented at annual meeting of Association for Education in Journalism and Mass Communication, Chicago.
Radford, T. 2003, 29 April. "Brave new world or miniature menace? Why Charles fears grey goo nightmare." *The Guardian*. Retrieved from www.guardian.co.uk/science/2003/apr/29/nanotechnology.science.
Redelmeier, D.A., and R.J. Tibshirani. 1997. Association between cellular-telephone calls and motor vehicle collisions. *The New England Journal of Medicine* 336(7): 453–458.
Rogers, E.M. 2003. *Diffusion of Innovations*. Glencoe, IL: Free Press, Macmillan.
Schiller, H.I. 1976. *Communication and Cultural Domination*. White Plains, NY: International Arts and Sciences Press.
Shatkin, J.A. 2008. *Nanotechnology: Health and Environmental Risks*. Boca Raton, FL: CRC Press, Taylor & Francis.
Sturgis, P.J., and N. Allum. 2004. Science in society: re-evaluating the deficit model of public attitudes. *Public Understanding of Science* 13(1): 55–75.
Takagi, A., A. Hirose, T. Nishimura, N. Fukumori, A. Ogata, N. Ohashi, S. Kitajima, and J. Kanno. 2008. Induction of mesothelioma in p53+/− mouse by intraperitoneal application of multi-wall carbon nanotube. *Journal of Toxicological Sciences* 33: 105–116.
Von Bubnoff, A. 2006. "Study shows no nano in Magic Nano, the German product recalled for causing breathing problems." *Small Times*. Retrieved from http://www.electroiq.com/index/display/semiconductors-article-display/270664/articles/small-times/environment/2006/05/study-shows-no-nano-in-magic-nano-the-german-product-recalled-for-causing-breathing-problems.html.
Ziman, J. 1991. Public understanding of science. *Science, Technology and Human Values* 16(1): 99.

3

Risk Communication in Theory and Practice

Risk communication practice is a professional subspecialty closely connected to fields such as risk analysis and risk management, risk communication research, and professional communication practice. Risk communication is concerned with identifying the most appropriate and effective ways of discussing risks with a broad variety of publics. Risk communication is also, in part, a research area often guided by—and, in turn, inspires—the study of how risks are perceived by various audiences; that is why this book is concerned with risk perception and risk communication taken together. Risk communication as a field of professional practice makes use of the knowledge gleaned from both risk communication research and risk perception research, alongside both general principles of good communication practice and an awareness of the ethical, political, and social dimensions of discussions of risk.

This subspecialty within communication generally embodies a philosophy of communication as a two-way process that takes audience perspectives and concerns into account. Risk communication research makes use of concepts such as risk information seeking and information processing (Griffin et al. 1999). It is concerned with understanding how audiences might respond to different forms of authority (Priest 2008), as well as to different messages. As social scientists, effective risk communication researchers usually do not begin with the assumption that public fears and concerns are necessarily irrational, even if these views might diverge from the views of experts. Much of the divergence in a given case likely reflects differences in underlying values and priorities, not necessarily disputes over scientific facts, as *deficit model* thinking might suggest. Even though accurate communication of the best scientific information is essential fodder for constructive dialogue, it is not generally sufficient to change or even to predict reactions that have their roots in deeper issues of trust (Priest et al. 2003; Siegrist et al. 2007) or in the existence of divergent social and cultural values, rather than purely scientific understanding.

Under crisis conditions, risk communicators are often primarily concerned with avoiding both rumors and the panic that rumors can fuel. This branch of risk communication work is often called *crisis communication*. In situations of crisis, institutional reputations are also often at stake. The management of the Johnson & Johnson Tylenol® recall in 1982, following several deaths attributed to cyanide poisoning from adulterated capsules, is often held up as a model for effective crisis communication that helped maintain the company's reputation through quick action and transparency in acknowledging the events and removing the product from store shelves. The near-disaster at the Three Mile Island nuclear power plant (Middletown, Pennsylvania) in 1979 is arguably even more famous as a case of bungled crisis communication, because reaction was slow and a variety of conflicting accounts circulated, with no one authoritative point of contact identified. These two cases and many others have resulted in several standard recommendations for crisis communication situations, including the advice to acknowledge problems and accept responsibility for them and the recommendation that institutions responsible for managing potentially risky products and facilities identify a single point of contact for responding to media and other inquiries.

However, outside of crisis situations, when there is more opportunity for people to be reflective, public communication about risks can follow the advice to be more interactive, ideally engaging risk managers, risk communicators, journalists, regulators, and members of broader concerned publics in constructive deliberation about how society might best approach particular risks. In part, this helps prepare everyone involved to react appropriately should a crisis subsequently occur. This approach is also consistent with a *marketplace of ideas* philosophy of the role of information in society. Yet some scholars reject the marketplace model. Power differentials among different groups in society tend to persist, discussion will not always eliminate conflicts among them, and political and policy developments will continue to be driven by the interests of the more powerful players. Risk communication is one area where these dynamics are often clearly visible. As the environmental justice movement has pointed out, risks are borne disproportionately by the poor rather than the rich, by minority rather than majority populations, and by workers rather than owners and managers.

Risk Communication Challenges

Risk communicators thus face many unique challenges. What constitutes the "correct" understanding of a risk, or of the degree of "riskiness" posed by a given hazard (e.g., the existence of a technology, or a group of technologies such as nanotechnology, that involves some level of risk), inevitably depends on factors that engineers may find difficult to factor into formal models, such as uncertainty, familiarity, or the number of people that might be affected (Fischhoff 1978). Some observers suggest that risk is actually inseparable from the perception of that risk, or, in other words, that the "actual" level of risk cannot always be clearly defined and neatly separated from what is perceived to be the level of risk (Bradbury 1989). Understanding this view can be instructive for risk communicators. This is not meant to sound mystical or to discount the existence of known risks that are both very real and very concrete. However, many risks are uncertain, and equally intelligent people from different walks of life bring different values (as well as different types of knowledge and experience) to the table in responding to these uncertainties. It is not always possible to identify a "correct" analysis of a given risk; even experts may disagree. In addition, neither the risks nor the benefits of technologies are always equally or fairly distributed across the population.

For example, for nanotechnology, the risks created over time by ongoing releases of nanoparticles of silver from wash water into ecosystems by nano-enabled clothes washers or nano-impregnated clothing, or the risks to which laboratory or factory workers might be exposed (via either inhalation or direct contact) involving the possibility of carbon nanotubes concentrating in their lungs or other tissues, are simply not yet understood. In both of these examples, the risks in question are still novel and uncertain, but they are nevertheless very real, whatever their magnitude is eventually judged to be. However, risk communicators must consider not only the scientific dimensions of these risks but also the political and psychological dimensions as well, in an environment in which the two types of risk cannot always be clearly separated from one another. For one thing, different groups will ultimately reach different judgments. Those with strong environmental values are almost certainly going to respond differently to a risk of environmental degradation than others will. And different groups also have different levels of both knowledge and power. Those who work directly with potentially risky nanomaterials may or may not be knowledgeable about those potentialities; they may also lack political power.

When members of various broader publics become concerned about a risk, especially one that is not well understood, it is usually more constructive (from a risk communication point of view) to listen to and address those concerns rather than make quick judgments and pronouncements about whose view is more rational. This is part of what is meant by a two-way approach. So far, most research on public receptivity to nanotechnology-based materials and products has found nonexpert publics (as well as expert publics) to be more or less unconcerned about potential risks. Even so, scientific research about these risks is ongoing, and it seems highly probable that the future will bring many situations in which society will be confronted with decisions about the risks of nanotechnology. Risk communication generally seeks to make this kind of confrontation as informed and constructive a process as possible; paralyzing polarization tends to serve no one's interests. At the same time, there are numerous situations when risk communication involves calling public attention to risks that have gone unrecognized. This is a difficult balancing act.

Moving forward into this uncertain future, consideration of nanotechnology's risks from a risk communication point of view will also involve recognizing that risk experts (often trained as scientists) and other audiences or publics are simply unlikely to understand risks the same way. While expert risk analysts generally try to distinguish between a *hazard* (something that could potentially be harmful, such as a sliver of broken glass on a kitchen floor) and a *risk* (the likelihood that the harmful potential will be realized in a given case, such as when an individual steps on that sliver of glass), risk communicators usually encounter audiences whose perception of a risk is strongly influenced by the characteristics of the hazard that is involved. In other words, the analytical distinction between risks and hazards may be of little help to the practicing risk communicator concerned with public reactions. Broader publics (generally those not trained in risk analysis) inevitably respond in ways that encapsulate emotional as well as cognitive or analytical aspects of risk evaluation. This is not to say that these broader publics are in any way "irrational" in their reaction to risks; rather, it is simply to recognize that their values, beliefs, and norms, alongside their emotions, also come into play.

My risk of being harmed or even dying in an automobile accident in modern society is real, concrete, and nontrivial, but it is a very familiar risk and may not therefore evoke the same level of day-to-day anxiety as a very small risk that an unfamiliar substance will turn out to be toxic. As a familiar risk, and one perceived as largely under a driver's control, the risk of death by car accident may tend to be taken for granted. Further, I need my car to get to work every day, so there are substantial and important barriers to my giving up this technology, or even to my fully confronting its riskiness. The very tangible necessity that I get to work, on top of the taken-for-granted character of automobile risks, diminishes my reaction to the risk, even though the risk in question is quite real. My risk of contracting an exotic fatal disease like ebola in my comfortable North American home is vanishingly small, but it can worry me a lot more than the much more likely risk of contracting fatal complications from infection by an ordinary influenza virus, because this latter risk is a relatively familiar one. Confronted with news accounts of a variety of risks every day, we all tend to perform a sort of *cognitive triage* that helps us evaluate which risks are important, and which can relatively safely be ignored. But here again our values and emotions inevitably come into play. (For more on the psychological factors that determine perceived risk, see Slovic 1987 and Siegrist 2010.)

Further, issues of values, not science, will determine whether a risk that affects only a few people severely (such as a rare hereditary condition) should be the object of the same level of societal concern, and research investment, as a risk that affects many people mildly (such as the common cold). A free-market economy provides more of an incentive for developing a cure for the common cold (which many people would buy) than for developing a treatment for a rare disease (marketable only to a few). Our consciences would generally recoil at the idea that we should give up searching for cures for rare diseases, and yet we are reasonably comfortable breaking the speed limit (when automobile accidents kill 1.2 million people around the world annually (WHO 2004) and generally ignoring the death toll from ordinary and highly treatable diseases (such as intestinal infections, which kill 1.5 million children under 5 years old annually) (UNICEF/WHO 2009). In the medical arena, patient advocacy groups play a special role in calling public attention to some diseases (such as breast cancer, in recent years), but perhaps not others that might affect equal numbers of people. Risk managers and risk communicators confront contradictions of this kind daily. Scientific logic is not the only arbiter of risk. Yet neither driving cars nor being disproportionately concerned about rare and unfamiliar diseases should necessarily be dismissed as irrational behavior, because both have other roots. We may have no access to alternative transportation, and we may be justifiably more concerned about a rare disease if the likelihood that it will spread has not been established.

One contemporary risk communication challenge involves parents who fail to get their children vaccinated against familiar childhood diseases because they have heard claims that these vaccines may cause autism. Present evidence suggests these parents are almost certainly making a mistake, but they are probably following an internal logic that makes sense to them. Perhaps they have lost their trust in the medical community because they feel that information about previous risks, such as the risk of contracting illness from animals infected with "mad cow" disease, were downplayed (especially in the United Kingdom). Probably they are more afraid of the relatively unusual and apparently mysterious condition called autism than they are of the relatively more common, familiar, and well-understood childhood diseases like measles and mumps (despite the fact that these are diseases that can kill—in 2008 there were 164,000 deaths from measles worldwide, see WHO 2009). Are these individuals best thought of as being irrational in these circumstances, or as only trying to do their best jobs as parents navigating a complex world in which not all information and not all experts can be trusted?

Could better risk communication resolve the vaccination–autism controversy? Could it have prevented it? Risk communication is not an exact science, so we can only pose these potentialities as questions. Current concern over declining childhood vaccination rates helps illustrate that attitudes toward new technologies are not determined exclusively by rational considerations. Yet addressing this issue will be most successful if it is respectful of parents' concerns, whether (on the surface of things) rational or not. Like these parents, most of us make choices daily out of perceived necessity and relying on imperfect and incomplete information; this does not make us irrational. And our existing sociotechnical system strongly constrains our choices in many areas, transportation being one of them. These kinds of contradictions pervade many risk communication situations, from both researchers' and practitioners' points of view.

The Goals of Risk Communication

Risk communication in a pluralistic democratic society obviously presents a variety of ongoing tensions and contradictions, including the coexistence of seemingly incompatible goals. On the one hand, most risk communication specialists would probably agree that their goal is not to engage in propagandistic manipulation of audiences, but to empower people to make good decisions based on those audiences' own values and beliefs. On the other hand, many risk communication practitioners work for institutions (ranging from research institutions to high-technology corporations to consumer advocacy organizations) that have powerful motives to persuade people of a particular point of view.

While a marketplace of ideas model proposes that public discussion can help reconcile these views—indeed, that the discussion is a sign of a healthy democracy—others are not so optimistic on this point, because some of these views are promoted by groups that have more resources available to them than do others. Risk communicators as individuals may be persuaded, perhaps with very good reason, that some perspectives on risk are right, while others are wrong. Just because society's reactions to particular risks have complex social and psychological roots does not mean those risks should be ignored or exaggerated. Further, risk communicators may work for employers with strong stakes in influencing public opinion in some particular direction. It is hoped that risk communication practitioners will approach these many contradictions with an awareness that it is often societal values and organizational interests, not simply scientific facticity, that are most centrally involved.

In short, some societal actors promote risk communication in the hopes of enhancing the democratic governance of technology—that is, of empowering citizens to make informed choices about which technologies should be adopted and under what conditions—while others seek to sell a product, to promote an idea, to raise levels of public concern, or to engineer acceptance (or rejection) of a technology in which they have a stake. These competing goals and assumptions create ongoing tensions. Risk communicators need to be aware of these tensions, which are often immediately relevant to their work. As is true for other types of messages, the *framing* of risk messages can sometimes have strategic goals. Framing, the choice of what information to put in and what to leave out in creating a message (Entman 1993), can be consciously designed to produce a particular effect (Snow et al. 1986). The framing of messages about nanotechnology's risks and benefits will be no different.

Democratic societies are generally based on the marketplace of ideas assumption that a free exchange of information and ideas will provide their citizens with the ability to make good choices about what information and ideas to accept and what to reject. The ability of any one actor to control the entire landscape (or marketplace) of ideas with respect to a given issue, be it about a technology or anything else, for an indefinite period of time is limited. The most effective messages are built on a strong foundation of both trust and transparency. So there is some reasonable hope that democratic ideals may win out, over time. Nanotechnology's emergence has provided the occasion for extensive experimentation with new public engagement strategies that is hoped will both encourage the involvement of various publics to participate in decisions about the management of technology and provide them with honest risk information, alongside information about benefits.

We do not yet know how effective these new approaches will be, but we do know that a strategy of avoiding discussions of risks will not be useful for long, in most cases. The existence of risks needs to be confronted head on; in fact, many individuals will likely react cynically to promotional efforts that are one-sidedly positive. Although public engagement and public discussion are not going to make risks go away, encouraging greater public knowledge and awareness regarding this new technological frontier is nevertheless good practice, for both technology and democracy. The benefits of new technologies will be more difficult to realize in an atmosphere in which risk communication is not part of risk management.

Risk communicators must walk a fine line between overstating and understating risks. Historical cases suggest that open and transparent acknowledgment of both risks and uncertainties is the wisest course of action. Risk communicators have ethical obligations to their profession and to society at large, even though they generally work for (and represent the interests of) employers and clients who are stakeholders for the issues about which they are communicating, and who may therefore have strong opinions about how to manage and communicate any risks involved. Further, risk communicators generally interact with a variety of publics who bring diverse attitudes, beliefs, and values to their interpretation of information and messages.

Best practices for risk communication involve encouraging two-way communication between these groups (that is, clients or employers on the one hand, and members of external publics on the other); in other words, a successful risk communicator brings the concerns of the members of broader publics to the attention of those who employ them, as well as provides information to those publics on their employer's behalf. They may play a role in shaping policy, not just transmitting it to others. And while, for science-related issues such as nanotechnology, part of the job of the risk communicator may involve encouraging public understanding of the underlying science, scientific knowledge does not generally shape attitudes or opinions; issues of trust and confidence, perceptions of justice or fairness, and underlying values, beliefs, and expectations also matter.

A successful risk communicator can therefore be seen (in part) as a negotiator among differing perspectives that may conflict, and this can clearly be a difficult role. It is made somewhat easier by a good understanding of how information about the risks and benefits of technology travels through society; what factors might cause individuals to accept and adopt or reject new technologies, products, and ideas; and how a particular group of technologies (in this case, nanotechnology) evokes particular opinions and reactions. A group of interrelated social theories (sometimes referred to as social theories of risk) help explain both the diffusion of information and the diffusion of new technologies. In addition, understanding the basics of opinion formation and how social scientists try to measure and explain opinions can also help. Some of the key theories are presented below.

Social Theories of Risk

Two theories already introduced briefly, diffusion-adoption theory (Rogers 2003) and social amplification theory (Kasperson et al. 2003), go a long way toward explaining societal reactions to new technologies of all sorts. Diffusion-adoption theory suggests that a variety of factors, including personality variables, as well as knowledge and the opportunity to try out something new, may influence individuals' decisions to adopt or to reject various technologies. Diffusion theory has been used by market researchers to identify and describe the various steps individuals go through when deciding whether to purchase and use a new commercial product, from initial awareness through trial of the product and then, finally, adoption. Naturally, this has been a very appealing and useful theory to these researchers, along with advertisers, manufacturers, and anyone else wanting to sell consumer products successfully.

Advertisers and marketers also seem to have been persuaded that labeling something *nanotechnology*—from the well-known iPod nano® to new fabrics, containers, cleaners, and a host of other products and devices—seems to leave a positive rather than a negative impression; at least, so it appears given the many marketing campaigns that have adopted this term. While the available evidence suggests that many people remain largely unaware of nanotechnology or have only a very general impression about what it might be, the word is already popping up in advertising for a broad range of items. It makes sense that in societies and cultures with highly developed, technology-based economies, and where new and innovative technologies are generally embraced with enthusiasm, nanotechnology might be quickly embraced as well. However, a positive reaction was not actually predictable given the controversies that have arisen in these same societies over other technologies such as nuclear power, genetically modified (GM) food, or stem cell research. Observers continue to be concerned that public opinion about nanotechnology could quickly turn negative as new information about potential risks is diffused, or as nanotechnology is fully incorporated into more sensitive products such as food packaging, as well as into agriculture. So far, however, this does not seem to have happened to the degree some have expected.

The bottom line is that we clearly still have much to learn about the adoption (or rejection) of technology by society. Social research has not yet uncovered all of the reasons that some technologies may be embraced while others are rejected in particular social, political, and cultural contexts. Reactions to risk are clearly socially patterned, not just individual, phenomena, however. Societal responses to specific new technologies are examples of what sociologists call *collective behavior*, something that involves processes that operate on a larger scale than the psychology of individuals. Individuals and groups influence one another, creating a *climate of opinion* on a certain issue that may be approving, disapproving, mixed, or indifferent. In turn, an individual's perceptions of what others are thinking can then reflexively influence that individual in turn, creating a sort of feedback loop in which opinions and attitudes can be reinforced or, in some cases, suppressed. This latter outcome is sometimes referred to as a *spiral of silence* and can result in particular opinions becoming less and less visible within society, especially when they are perceived to be minority opinions (Noelle-Neumann 1993). Crowd behavior is another example of a collective social process, as are the emergence and spread of rumors and urban legends. Underlying all of these processes are the values and beliefs that are broadly shared among members of a particular cultural group.[1] Theories of the relationship between such cultural factors and reactions to risk are sometimes called *cultural theories of risk* (see, e.g., Douglas and Wildavsky 1982).

Arguably, while market research designed to support product sales most often seeks to understand consumer behavior on the individual level, diffusion theory and related social theories of risk are most powerful when they are applied at the cultural or societal level. This is because diffusion is an inherently social (that is, collective) process, a process that takes place only in groups as information about new technologies and products diffuses through social networks. This theory first became well known among applied social researchers for its value in agricultural extension work, providing a useful explanation of why farmers may or may not adopt a particular new agricultural technology. This same idea has also been used by archaeologists to trace the spread of a new type of arrowhead or other tool among prehistoric humans, across both time and geography. Individuals are particularly influenced by respected others (in an agricultural context, a neighbor who is a successful farmer or a local community leader) in a process known as opinion leadership (Katz and Lazarsfeld 2006), a process probably as important now as it was 10,000 or more years ago.

Opinion leaders may be respected parents and elders, teachers and other educators, counselors, religious or political authorities, and so on; they may be the young and innovative people who tend to set new trends or more established, prestigious individuals who have recognized expertise in a particular area, such as a doctor or an engineer. We each tend to have different opinion leaders in different areas of our lives; for example, I may have one friend who is a film aficionado and guides my choices as to what movies to see, and another, quite different acquaintance whose political ideas I trust and whose opinions may therefore affect where I stand on social issues, perhaps even influencing my voting decisions. In recent years, this concept of opinion leadership has also been utilized by market researchers, alongside diffusion theory more generally, as these researchers seek to identify those early adopter shoppers who may be most likely to try an unfamiliar product and later influence others who follow their lead. Opinion leadership, like diffusion, is a concept that operates at the societal level; an individual living alone on a proverbial desert island cannot be an opinion leader, nor can he or she (as an isolated individual) be influenced by opinion leaders.

[1] The words *culture* and *society* describe groups of human beings, of course. But the terms *culture* and *cultural* draw attention to shared values and beliefs, while the term *society* or reference to social phenomena emphasizes the way that the group is organized (e.g., its institutions and the social roles people occupy within them).

Highly relevant to the formation of opinions about new technologies, both opinion leadership and associated patterns of trust in what particular types of authorities have to say are phenomena that vary from culture to culture (Priest et al. 2003; Priest 2008), and these patterns undoubtedly influence adoption patterns. The influence of opinion leaders may very often be most powerful when it is interpersonal, taking place when individuals interact face-to-face. However, these same dynamics will affect how media messages such as news stories or advertisements are interpreted. Such messages, both news and ads, are commonly delivered by friendly and smiling, attractive, authoritative, and sometimes celebrity faces and voices. These media personalities often engage their readers and viewers in a way that very much resembles (psychologically) ordinary, face-to-face, forms of interpersonal communication. So when we think about the dynamics of diffusion theory, we should not just imagine lone individuals as they go about their daily lives as individuals, viewing advertisements, browsing the Internet, reading news stories, and so on, but we should also think about the patterns of interpersonal and group conversation, interaction, and persuasion in which that person participates, including the vicarious "interactions" they have with media personalities and the ways that they are affected by these.

The spread of information about the risks and benefits of technology (as about many other matters) through social networks, including media networks, has received an enormous amount of attention from social researchers. A clearly important function of society is to provide mechanisms to alert its members to potential harms, a role that has undoubtedly been important since prehistoric times, when we can readily imagine the news of threat from fires, floods, storms, predators, or human enemies surging through small bands of hunters and gatherers. Risk communicators are the direct descendents of the members of these early human groups who stood watch on hillsides and spread the alarm when trouble was spotted coming over the horizon. Yet no one wants society to panic unnecessarily, to repeat the pattern associated with Orson Welles' 1938 "War of the Worlds" broadcast, when some early radio listeners took a science fiction story about aliens from outer space invading the East Coast at face value, and some of them panicked.

This event has now become an almost apocryphal example of overreaction to news of an imagined risk delivered by the media. Not everyone who heard this broadcast became alarmed. Ironically, in truth, sociologists studying the "War of the Worlds" events found that most people did not panic, and those that did often belonged to groups that can be characterized as especially susceptible to being influenced (Cantril 2005). It is interesting how much attention, even today, is given to remembering those who panicked, and how little to considering those who remained calm. In truth, the risks that society initially ignores—for example, many long-term risks to health and to the environment, or the risks associated with radiation, deep water ocean drilling for oil, or ordinary automobile travel (especially in the days before seatbelts, infant car seats, and airbags), or the risks that the flood control system in New Orleans would fail, inundating much of the city—turn out to be all too real. How does this happen, that some imaginary risks get so much attention, while other very real risks get so little?

Social amplification theory deals both with why some risks become exaggerated (or amplified) and why others become ignored or de-emphasized (attenuated), as information about those risks travels through social networks, including (but certainly not limited to) the appearance of that information in mass communication messages in various forms. However, while social amplification theory is concerned with both the social amplification and the social attenuation of risk, it is almost always the amplification of risk that seems to generate the most attention and concern. Perhaps this is simply because situations in which risks appear (rightly or wrongly) to have become exaggerated are just more visible, more public, than situations in which risks are sidelined or ignored (or, in some cases, not even discovered until too late). The emergence of protestors, news stories, watercooler conversations, political speeches, and so on generally follows the path of information about the existence of risks, not their nonexistence. For whatever reasons, and even though in actuality risks are not always amplified (despite how things might appear on the surface sometimes), those situations in which risks do seem to be amplified certainly get the lion's share of attention from politicians, policy makers, business leaders, news organizations, and even scientists and engineers, and, ultimately, of the social scientists and risk communicators called on to better understand people's reactions. As the managers of our "risk society," all these groups constitute the individuals who must cope with society's concerns, whether exaggerated or not.

Society's risk managers, in addition to being concerned on behalf of society generally, may be concerned about their own vulnerabilities to risk in a somewhat different way. If risks are seen as out of control, politicians may not get reelected, regulators may be called on the carpet and urged to take hasty action, scientists and engineers may fear their research and development funding will be diminished, and the marketers of new technology may face unreasoned consumer rejection. Further, business leaders may worry that both market uncertainty (the issue of whether people will accept their products) and regulatory uncertainty (the issue of whether new laws and regulations will suddenly change what they can and cannot do) could create additional management problems for them when risks become amplified.

The interests of all of these powerful groups are threatened more immediately by risk amplification than by risk attenuation. So they turn to risk communication specialists to help them out. In reality, of course, attenuated risks are also sources of concern. Businesses do not want to market products that create unexpected liabilities for them, the medical community does not want people to ignore the risks of disease, accident, or invisible but health-threatening pollution, and so on. These situations simply get less attention a good deal of the time, which is why risk communicators (especially those employed by government or by advocacy or watchdog groups) sometimes find themselves in the position of calling more attention to little-known risks, rather than always working to get known risks accepted. At present, both the benefits and the risks of nanotechnology are receiving limited attention, as opposed to what historically happened with the most commonly cited comparative example, the emergence of popular resistance to genetically modified food crops. While social amplification theory does not yet do a very good job of predicting which risks will become amplified and which ones attenuated, this comparison yields some important insights.

PERSPECTIVE: THE NATIONAL RESEARCH COUNCIL'S 1989 RECOMMENDATIONS FOR IMPROVING RISK COMMUNICATION: RISK COMMUNICATION AS AN INTERACTIVE PROCESS

Thomas B. Lane, Jr.

In 1989, the National Research Council issued a report that addressed the increasing difficulty risk communicators, in particular those in government and industry, had been facing in successfully communicating accurate, clear, and straightforward messages about hazards and their associated risks (Committee on Risk Perception and Communication 1989). The report suggests that previous research into risk communication has suffered from two fundamental weaknesses. First, the focus has been on one-way communication from experts to nonexperts. While primarily concerned with risk messages coming from government agencies and private industry to nonexpert recipients, the report also makes specific recommendations that seek to solve the problems faced by those recipients. The second weakness the report addresses was that previous research had typically placed emphasis on the content of specific risk messages, neglecting the risk communication process as a whole (2).

This influential report stressed the importance of viewing risk communication as an interactive process by which individuals, groups, and institutions exchange information (2). Thus, successful risk communication occurs not simply when the communicator persuades the recipients of a particular message to accept the intended positions or recommendations, but, rather, risk communication is successful only insofar as it adequately informs all decision makers, including government officials, industry managers, and individual citizens, of relevant issues or actions pertaining to some particular risk. In addition, the report suggests, successful risk communication assures these decision makers that they are informed to the greatest extent possible given the best available knowledge (2–8).

Certainly, however, there are situations in which the goal of risk messages may reasonably go beyond simply informing, such as in cases where significant public health risks are involved. One particular concern raised in the report was the extent to which a public official ought to attempt to influence individual decisions regarding a risk in cases where doing so seems justified. The report recommends that when situations arise that necessitate messages that are more persuasive, public officials should legitimize such actions via a democratic process, such as by relying on some form of independent review (2–3). Yet, while such approaches might serve to legitimize the persuasion sometimes used in risk communication, the authors pointed to problems endemic to the institutions in which the practice of risk communication is embedded. Legal and political issues might have an impact on both the content of risk messages and how they are disseminated, and they also may impose certain limitations on what action interested parties may take when confronted with a risk (5).

The report also argues that effective risk communication could serve to increase citizen participation in a democracy. For example, increased awareness of particular risks might motivate citizens to attempt to exert more control in the decision-making process. However, decision makers at the local level are less likely to take into consideration the interests of the wider public (5). In addition, the report suggests that the interests of risk communicators and the groups they represent, as well as their own beliefs and predispositions, can produce bias that, in turn, leads to inaccurate or misleading information. Further, if the risk communicator relies on a source that a particular group does not trust in addressing its concerns, the group may reject the information as a basis for decision making (5–6).

The sense that a particular group's interests are not being adequately addressed by the source of a risk message is not the only factor that affects whether decision makers rely on the message's content. The report also identifies a lack of credibility on the part of the risk communicator in the eyes of the target audience as hindering successful risk communication. This lack of credibility leads to distrust and hostility. The report finds that an important factor affecting the credibility of both a source and its messages is the perception by the message recipients of the legitimacy of the process that determined the contents of a risk message. These perceptions of the legitimacy of the process depend upon several factors, including the legal standing of the source, the justification provided for the communication program, the extent to which affected parties can participate in the decision-making process, and whether conflicting claims are reviewed in a fair and balanced manner (6–7).

The report also notes that message recipients' perception of the accuracy of messages affects (in turn) their perception of the source's credibility. They found recipients are less likely to trust the accuracy of the message if they perceive that the message advocates a position that is inconsistent "with a careful assessment of the facts" or if a source has "a reputation for deceit, misrepresentation, or coercion" (6). Message recipients may doubt the accuracy of a message if it conflicts with previous positions taken by the same source or with messages from other, more credible sources. Finally, the authors found, recipients might likely perceive a message as inaccurate if its content seems self-serving or if its source is perceived as professionally incompetent (6–7).

The report also addresses the difficulty risk communicators face in presenting the scientific and technical information necessary in accurate risk communication in a manner that uses concepts and language easily accessible to ordinary people. Moreover, they note that risk communicators must sometimes communicate something even though there may not be enough information to support a particular recommendation conclusively, such as in cases of emergencies requiring immediate action or when there is a pressing need for the reporting of the preliminary findings of a study or analysis (7).

In addition, risk communicators often encounter difficulty getting the attention of the intended recipients and getting those recipients to address pertinent issues (7). Sometimes recipients may have difficulty deciding which issues are important because the messages they receive do not provide them with information that they feel satisfactorily addresses their concerns or the information they receive is not from a trusted source (7–8). Even when risk communication is successful, the result may not always be what a particular source intends.

Likewise, successful risk communication need not bring about consensus when addressing controversial issues. Because interests and values vary, the report argues, better understanding of a particular risk might lead individuals to draw different conclusions about that risk. Thus, rather than merely improving the content of risk messages, not only must risk communication procedures be improved, but also risk management and risk analysis approaches may need to be reconsidered (8).

The report makes recommendations designed to improve the content of risk messages and the management of the risk communication process and to increase recipient participation and understanding. Overall, the report recommends that risk messages should be consciously designed to meet the practical goals of improving understanding of issues, while also explicitly considering the roles of potential recipients within the political and legal context of the risk communication process (9). Moreover, risk communicators should engage in an early, sustained, and open dialogue with potentially affected individuals and groups, avoid relying on briefings restricted to technical issues, and explicitly define the limits of outsider participation (9).

Recommendations aimed at risk managers also addressed the need to ensure that risk messages will not become distorted. Risk managers should hold those responsible for generating risk messages accountable for detecting and reducing distortion, and when feasible, they should consider independent review of the risk assessment and message by recognized experts, allowing outside preview of draft messages and encouraging "white papers" on related issues (10).

The report also recommends that risk managers should carefully consider the composition of the intended audience and how well the message and its dissemination address the audience's concerns. They should rely on communications specialists, as well as risk experts, to produce accurate assessments and the right risk message. Risk managers should develop a program to evaluate the success of previous risk communication, which also should serve to improve their organizations' understanding of the roles and limitations of intermediaries, particularly the media (10–11). In cases where it is foreseeable that an emergency can arise, risk managers should have advance plans for communication in place, including the coordination between various authorities and a centralized outlet where the public and the media can obtain access to reliable and current information (11).

REFERENCE

Committee on Risk Perception and Communication, National Research Council. 1989. *Improving Risk Communication.* Washington, DC: National Academy Press.

References

Bradbury, J.A. 1989. The policy implications of differing concepts of risk. *Science, Technology & Human Values* 14(4): 380–399.

Cantril, H. 2005. *The Invasion from Mars: A Study in the Psychology of Panic.* Piscataway, NJ: Transaction.

Douglas, M., and A. Wildavsky. 1982. *Risk and Culture: An Essay on the Selection of Technical and Environmental Dangers.* Los Angeles: University of California Press.

Entman, R.M. 1993. Framing: toward clarification of a fractured paradigm. *Journal of Communication* 43(4): 51–58.

Fischhoff, B., P. Slovic, S. Lichtenstein, S. Read, and B. Combs. 1978. How safe is safe enough? A psychometric study of attitudes towards technological risks and benefits. *Policy Sciences* 9(2): 127–152.

Griffin, R.J., S. Dunwoody, and K. Neuwirth. 1999. Proposed model of the relationship of risk information seeking and processing to the development of preventive behaviors. *Environmental Research* 80(2): S230–S245.

Kasperson, J.X., R.E. Kasperson, N. Pidgeon, and P. Slovic. 2003. The social amplification of risk: assessing fifteen years of research and theory. In *The Social Amplification of Risk.* Eds. N. Pidgeon, R.E. Kasperson, and P. Slovic. Cambridge: Cambridge University Press, 13–46.

Katz, E., and P.F. Lazarsfeld. 2006. *Personal Influence: The Part Played by People in the Flow of Mass Communications.* New Brunswick, NJ: Transaction.

Noelle-Neumann, E. 1993. *The Spiral of Silence: Public Opinion, Our Social Skin.* Chicago: University of Chicago Press.

Priest, S.H. 2008. North American audiences for news of emerging technologies: Canadian and US responses to bio- and nanotechnologies. *Journal of Risk Research* 11(7): 877–889.

Priest, S.H., H. Bonfadelli, and M. Rusanen. 2003. The "trust gap" hypothesis: predicting support for biotechnology across national cultures as a function of trust in actors. *Risk Analysis* 23(4): 751–766.

Rogers, E.M. 2003. *Diffusion of Innovations.* New York: Free Press.

Siegrist, M. 2010. Psychometric paradigm. In *Encyclopedia of Science and Technology Communication*, 2. Ed. S.H. Priest. Thousand Oaks, CA: Sage, 600–601.
Siegrist, M., T.C. Earle, and H. Gutscher. 2007. *Trust in Cooperative Risk Management: Uncertainty and Scepticism in the Public Mind*. London: Earthscan.
Slovic, P. 1987. Perception of risk. *Science* 236(4799): 280–285.
Snow, D.A., E.B. Rochford Jr., S.K. Worden, and R.D. Benford. 1986. Frame alignment processes, micromobilization, and movement participation. *American Sociological Review* 51(4): 464–481.
UNICEF/WHO (World Health Organization). 2009. "Diarrhoea: why children are still dying and what can be done." Retrieved from http://whqlibdoc.who.int/publications/2009/9789241598415_eng.pdf.
WHO (World Health Organization). 2009, December. "Measles." Retrieved from www.who.int/mediacentre/factsheets/fs286/en/.
WHO (World Health Organization). 2004, March 29. "Road safety: a public health issue." Retrieved from www.who.int/features/2004/road_safety/en/.

4

Public Opinion, Public Perception, and Public Understanding

Regardless of how much we think we understand about public reaction to the risks and benefits of technology, it is not always easy to distinguish, when a technology is first introduced, either how risky the technology might ultimately prove to be, what kinds of risks will emerge as most prominent, how these might be weighed (in public opinion) against perceived benefits, or how people are ultimately likely to respond. Although we have many clues from the research literature on this subject, it is never entirely obvious which risks society will react to most strongly. These reactions involve not only the psychology and previous experiences of individuals, but also the synergy and mutual influence among various segments of society, both groups and influential individuals. Genetically modified (GM) crops and food products, briefly introduced in Chapter 2, provide a handy recent example of how public opinion forms around perceived risk, as well as the ways that society's leaders and risk managers (in turn) tend to react to those perceptions.

At the time, anti-GM sentiment seemed to arise out of nowhere. The term *Frankenfoods*, which suggested a perception of GM as not only grossly unnatural but also frighteningly out of control, as well as a whole host of associated images, seemed to capture public sentiment, and at least to some observers, seemed to create irrational and unfounded public fears out of nothing. Protestors taking to the streets over the integrity of their food supply, a phenomenon that in the case of GM was probably more common in Europe than the United States but became visible around the world, even today,[1] was unprecedented and largely unanticipated. Early GM foods, such as a tomato designed to have a longer shelf life, had been marketed as a technological advance and adopted largely without incident, so why the sudden uproar? Media scholars would caution against assuming that use of a particular term or description could be responsible for such widespread action and concern, unless other elements were also in play.

[1] For example, anti-GM protestors marched outside the 2010 meeting of the American Association for the Advancement of Science in Santa Barbara, California, because of the presence of Monsanto Corporation, a major GM-producing corporation that was reported to be sponsoring refreshments for journalists covering the meeting.

Some scientists, policy makers, and industry representatives clearly felt that exaggerated risk information or poor understanding of the underlying science was entirely responsible for public objections to GM foods in both Europe and the United States, as well as questions about GM raised in other parts of the world, but alternative explanations are available. These explanations are consistent with understanding risk perception as deeply rooted in strongly held values. Right or wrong, some GM opponents feared that GM would further disadvantage the traditional family farm in the agricultural marketplace, for example, or otherwise damage a way of life they saw as worth protecting (Priest 2001). Some of the first public protests over rbST (recombinant bovine somatotropin; bovine growth hormone artificially produced via genetic engineering and fed to dairy cattle to boost production) arose in small-farm dairy states.

The GM Food Story Revisited

Although some observers might argue that Americans' reverence for the family farm should be seen as out of step with the current reality of agribusiness, it is clearly an attitude that persists. Other factors may have included fears that subtle environmental or health impacts had not been sufficiently researched, or the results broadly enough communicated, which may have helped undermine public trust in both the research process and scientific leadership, even in the absence of clear evidence of possible harm. Yet other objections to GM came from the organic foods community, who may have objected to interference with what they saw as a preferable, perhaps more natural, means of food production and an approach associated with a different form of social organization for agriculture that emphasized smaller farms. Even U.S. university faculty felt that the emphasis on development of GM distorted the agricultural research agenda (Priest and Gillespie 2000). Most of these concerns reflected values, priorities, preferences, and issues of trust, rather than either understanding or misunderstanding of scientific facts.

The concern over openness and transparency about safety-related issues as GM products moved onto farms and into markets may have been particularly acute in Europe in part because these products had initially been developed in the United States and were therefore perceived as representing foreign interests. This became an additional issue of values, one intertwined in complex ways with issues of national autonomy and of trade relationships across the Atlantic. Public opinion ultimately became polarized in the United States, however, not just in Europe or (somewhat later) in India and Africa. Publicity about rejection of GM elsewhere was quite visible in the United States—publicity that (ironically) put little emphasis on the resistance taking place within the United States, raising the question of to what extent there were really fewer objections in the United States or whether they simply received less media attention there.

With respect to the "developing" or "less-developed" world, there was particular concern with protecting the livelihoods of subsistence farmers if the local agriculture system was subject to major forms of change as a result of technology being introduced by outside interests whose purposes were often suspect to begin with. In a postcolonial era, the desire of multinational corporations to extract new sources of income through exporting new agricultural inputs (including patented seed) and exploiting other nations' genetic resources (such as plants with pharmaceutical or other marketable properties) for profit naturally caused concern. Parallels in the developed world include the significant concern with protecting the family farm in both Europe (where many such small-scale farms are very actively protected by government policy) and the United States (where the family farm has failed to pay most family's bills for generations, yet today remains the romantic ideal). In short, not much in the way of anti-GM reactions can be attributed to flawed scientific understanding of the risks in comparison to that attributable to issues of trust, economics, and societal values that were also in play.

Of course, GM foods and other crops certainly have significant potential to bring good as well as bad effects, and GM continues to be explored for its ability to produce more nutritious foods, including foods delivering otherwise scarce nutrients. One example is the much-touted "golden rice" that could provide consumers, especially rice-dependent Asian consumers, with a new source of the vitamin A precursors that can prevent blindness and even death for many. Other developing-world applications, some of which are the subject of ongoing research, include food crops adapted to marginal agricultural lands and those that (it is argued) can be grown with fewer pesticides and herbicides. Some of these claims remain rather controversial, with many of the potential benefits not fully realized (and much in the way of GM crops consisting of cotton, soy, and corn crops grown on very large-scale farms). Yet claims about benefits have appeared to some observers to be completely drowned out in a sea of concern over risks, with the potential benefits seemingly dismissed by some audiences as so much pro-GM propaganda.

Regardless of exactly where the truth lies, this kind of polarization is unlikely to result in constructive solutions, and again highlights the place of values in debates that on the surface might appear to be about "science." In dynamics parallel to those of the orphan drug dilemma, the problem also remains of who is going to fund the development of crops that are of primary value to economically marginal subsistence farmers and the local populations they feed.

A major lesson learned for risk communicators from the GM debates was that "deficit model" (see Chapter 2) thinking—that is, the assumption that misinformation and scientific illiteracy are largely the root cause of negative public reactions to new technology—was not adequate to explain public reactions to this controversy. The public reaction clearly has other foundations, ranging from failure to involve broader publics in discussions earlier in this technology's deployment to failure to understand the concerns about whether possible economic, environmental, and health effects were justly distributed or adequately understood, not to mention the perceived "unnaturalness" and danger of artificially altering DNA. The solution to risk amplification is not always simply more education about the relevant science, in other words. Popular objections to GM crops and to many other technological innovations generally reflect deeply held social values; this is as true today as when "Luddite" objections to the automated weaving technologies of the nineteenth century in the United Kingdom reflected the perceived potential for disruption of a way of life (Sale 1999).

Fears have often been expressed but are so far unrealized, for the most part, that nanotechnology will eventually go the way of some of these forms of biotechnology, with amplified fears about risks overshadowing popular acceptance of potential benefits. It is, of course, ironic here that risk may not always be what is primarily driving public concerns, but risk issues often seem to become the battleground; therefore, better risk communication is seen as a solution. Perhaps this is simply because (at least in the United States) we have a much stronger tradition of managing and regulating technologies based on their riskiness than for their anticipated economic and social consequences or in response to people's ethical objections. At any rate, so far, nanotechnology is being received in a largely positive climate of public opinion (Gaskell et al. 2005). Meanwhile, social scientists, as well as a variety of advocacy groups (including those interested in promoting nanotechnology's development, adoption, and diffusion), continue to keep an eye on emerging public opinion about these technologies, in the first case seeking primarily to advance our understanding of how public opinion about any emerging technology takes shape, and in the second hoping to get an early warning sign of shifts in the climate of public opinion.

It is within this historical context of largely unanticipated negative reaction to biotechnology (and, earlier, to the adoption of nuclear power, which inspired a previous generation of social and behavioral scientists to develop research programs directed at understanding risk perception issues) that risk communication surrounding nanotechnology is taking place. Today, much more attention is being paid to analyzing the reasons for risk perception and opinions about new technology. Yet public overreaction is still actively feared, and public ignorance is still regularly blamed for reactions that seem problematic, generally because risks are seen (often by those with stakes in the promotion of those technologies) as having been blown out of proportion.

Of course, in a modern society with a technology-based economy, science literacy is an absolutely essential component of citizenship, and it is sometimes woefully lacking. But most members of many of these societies are essentially pro-technology, and communication specialists embracing a two-way approach generally take a different stance, arguing that while it is important to explain and educate about the science underlying new technology, it is equally (or more) important to educate leaders and experts about what other segments of the public might think. This matters to efforts to maintain an atmosphere of trust and cooperation between policy makers and publics, to provide opportunities for broad public input into making decisions about technology adoption, and to address popular concerns as directly and meaningfully as possible. The policies that are adopted and the information and messages about them that are disseminated will meet a more positive public if reception is a result of two-way communication that does not necessarily dismiss popular views as springing from irrational amplification.

Opinion Studies and Their Implications

As a result of this general concern with public reactions, public opinion about nanotechnology has been assessed by a number of major polls in recent years, beginning as early as 2002, when an Internet-based survey of the opinions of 3909 English-speaking volunteer participants appeared in the *Journal of Nanoparticle Research* (Bainbridge 2002). This survey claimed to reveal "high levels of enthusiasm for the potential benefits of nanotechnology and little concern about possible dangers," although this conclusion was based on only two questions and targeted participants likely to be especially oriented toward science, representing the "attentive public" (Miller 1986) for science—that is, those with a special interest in scientific issues. (Two major scientific organizations sponsored the study, a factor likely influencing the profile of the participants.) Despite the almost certain attentive public skew, it is nevertheless worth noting that 57.5% of these respondents agreed that "Human beings will benefit greatly from nanotechnology." (Bainbridge 2002, 562)

Of course, the attentive public for science is only one of the many crosscutting and intersecting groups making up most contemporary societies. Even though it is possible to imagine that as our economy further globalizes, societies and cultures will become more homogenous, globalization has actually been accompanied by new patterns of migration, resulting more visibly in increasing cultural and ethnic diversity rather than increasing cultural homogenization. Many of today's societies are experiencing this increasing diversity, not just societies like the United States, which has historically been diverse as a result of the importation of African slaves and the successive waves of European and Asian immigrants. As a result, thinking in terms of any single *public* when talking about public opinion can be misleading. This is one reason why, in this book, we generally speak about *publics* rather than use the phrase *the public,* which brings to mind an undifferentiated mass (or *mass public*) made up of relatively similar people. In fact, in the diverse pluralistic societies of today's globalizing world, there simply is no one *general* public anymore—that is, if there ever was one.

Most opinion polls aim to generate results that describe the opinions of "people in general," and they do this by attempting to reach a random[2] sample of respondents. Because reactions to technology often involve values and beliefs, opinion polls can provide only limited information. A number of factors interfere with getting an accurate barometer that can measure public thinking. Random samples are likely to do a good job of reflecting the population in question, but they are increasingly difficult and expensive to create (due to factors such as increasing social mobility, meaning lists based on addresses or telephone numbers rapidly go out of date, as well as rising cell phone use, so that no one telephone system or listing is an adequate starting point). Lowered response rates, partly due to survey burnout, as we are all asked to complete numerous polls on a regular basis, also introduce bias. This is not limited to the issue of attracting people who have special interests in the topic of a given survey; responders may also have more cooperative personalities, more time on their hands, or have already formed strong opinions they want to express.

In the case of nanotechnology, it may be that most opinion polls reflect varying degrees of these biases among those willing to participate in a poll about such a seemingly obscure and technical topic. Further, question wording can and often does introduce other unintentional biases, which can be especially important for a complex new topic that is largely unfamiliar (Bishop 2005). Nevertheless, polls are a powerful technology and one of the best ones we have for assessing what broader publics are thinking at a snapshot in time.

[2] In this context, random means that each member of the population—or group of people under study, often the general public—has an equal chance of being accepted. This can rarely be achieved in practice, so a representative sample is the next best thing. In a representative sample, the proportion of people in various demographic groups such as age, gender, ethnicity, and so on, at least resembles the proportion in the population under study. Sometimes representativeness is achieved by quota sampling (recruiting a sufficient number of participants from each of a number of predetermined demographics).

So what does the public in general think about nanotechnology, exactly? And how do we know? Despite the caveats explained above, a number of other opinion polls and other recent studies have consistently documented the existence of a largely positive opinion climate in those areas of the world that have been studied by these methods. (Polling remains difficult in countries with large rural populations and lower levels of economic development, such as China and India.) These more recent surveys, using random samples and generally more sophisticated methods, have provided results headed in approximately the same direction as Bainbridge's 2002 study: The public opinion climate still seems to be generally a positive one for nanotechnology.

Overall, most available opinion polls have not only reflected little public concern about nanotechnology, but also in most cases, not even much familiarity. One meta-analysis of 11 other public opinion studies carried out in North America, Europe, and Japan and published between 2004 and 2007 demonstrated that large numbers of individuals stated they had no knowledge of nanotechnology (that is, they knew "nothing at all") (Satterfield et al. 2009). Eight of the 11 studies put the proportion of those unfamiliar at over 40%; the remaining three included the smallest numbers of participants and thus would have had the largest margins of error (meaning that the results of these smaller studies, while worth taking into consideration, are statistically less certain). Even though this suggests that public opinion is still malleable, there is thus far little evidence that risk amplification is taking place, with about a 3-to-1 ratio of those perceiving greater benefits to those perceiving greater risks, even though almost half (44%) remain unsure.

In light of the discussion above, it is especially interesting that the authors of this study expected to find a lack of familiarity to be associated with risk aversion, something predicted by previous research on other technologies, but this was not the case. They also found that more knowledge was associated with positive perceptions. Although at first glance this might seem to imply that as a person's knowledge of nanotechnology increases, the increase causes them to be more positive, it can also be explained in a different way. In an environment in which many people are largely unaware of a technology, those who are more aware are likely to be exactly those pro-technology early adopters who tend to embrace, rather than reject, technology—the same people who may well have been overrepresented in the earlier Bainbridge study. Scientists, engineers, and others whose work involves these or other technologies are certainly a part of this group, as are others with generally pro-technology attitudes. Absent evidence of public concern and sustained media attention to risks, and in a generally positive climate of public opinion, most people are not as risk-averse as these researchers had expected.

Another recent overview (Currall 2009) summarizes available public opinion studies, including two major studies that conclude that values and beliefs drive acceptance (Kahan et al. 2009); this overview also reports on a U.S.–U.K. comparison that the author argues helps underscore differences between the two cultures, with the United States reflecting more optimism about technology and the United Kingdom more awareness of societal implications at various levels. However, the same U.S.–U.K. study also reveals a shared emphasis across the two cultures on benefits, suggesting a generally positive perspective on science and technology in both countries (Pidgeon et al. 2008). This might not be entirely surprising, given the proportion of cultural history that is shared between these two countries, which of course also share a common language.

Much more striking is a report by Fujita, Yokoyama, and Abe (2006) that 55.2% of Japanese are aware of nanotechnology, and 88% are "hopeful" about its benefits, compared to other cited figures of 48% awareness and 70% optimism in the United States, and 29% awareness and 68% optimism in the United Kingdom. Many competing explanations might exist for observed cultural differences such as these, ranging from differences in sampling to differences in wording (an especially challenging issue across languages). Even so, the preponderance of available evidence stresses the primacy of culture-based values in shaping initial reactions to nanotechnology, with Japan seemingly coming out the most aware and the most optimistic of the three.

Surveys are sometimes referred to as *one-shot* research designs because the data are gathered from a broad spectrum of individuals at a single point in time. It is characteristic of this type of one-shot research that cause and effect are especially difficult to disentangle. For example, do people who are more aware of technology seek out more information, or does the incidental acquisition of information change attitudes? This can be difficult to untangle. Using other evidence from a 3-year panel study of 76 citizens in the southeastern United States, Priest, Greenhalgh, and Kramer (2010) found that over the first year, risk awareness did rise somewhat among both heavy media consumers and other participants. Further, while risk awareness rose in both groups, the heavy media consumers in the study both began and ended with higher levels of risk awareness, suggesting that mediated information may have made a difference (although it may also be, once again, that the heavy media consumers were simply more technology-aware to begin with).

However, overall, changes in risk perception were much less visible among those who completed the entire 3 years of the study. As is common in panel studies, which attempt to follow the same group of people over a considerable period of time, not everyone completed the study despite the availability of modest payment incentives for finishing; thus, while the one-shot problem is overcome, a new problem of participant attrition is introduced, and study completers are likely different from study dropouts. In this case, about half of the original participants remained for the entire 3 years, and these tended to be from higher education and income levels (Priest et al. 2011). The patterns discerned in this study, nevertheless, suggest the possibility that even among those attuned to early warning signs of technological risk, risk amplification does not always take place. That is, opinions are not always volatile, as some observers have assumed.

Implications for Risk Communication Research and Practice

As noted, exactly what causes risk amplification under some circumstances and risk attenuation under others as a new technology diffuses and is adopted throughout society remains imperfectly understood. Lessons learned from earlier technologies do not apply very neatly to the case of nanotechnology, which suggests that nanotechnology is different, perhaps that it does not have the same degree of "cultural resonance" (see below) as some other forms of technology, especially biotechnology. News media are considered important *amplification stations* in social amplification theory—one of the key institutions identified in social amplification theory (Kasperson et al. 2003). Yet, even though news accounts have picked up scientific reports that some forms of nanotechnology (specifically, carbon nanotubes) may behave like asbestos in the human body (Bergstein 2008; Poland et al. 2008), implying they may cause mesothelioma just like asbestos does, even this does not seem to have caused much in the way of widespread concern.

To social scientists studying risk, this is an intriguing conundrum. What is it that we thought we knew about risk perception that does not apply in the case of nanotechnology? Many speculate that as nanotechnology and biotechnology converge in novel therapies and agricultural applications, it could be that the present, largely positive, climate of public opinion will evaporate. In the future, perhaps we may even see nanotechnology-based technologies for enhancing humans, such as by increasing their memory capacity, as well as curing disease in new ways, perhaps by facilitating gene therapy. But it is impossible to predict public reaction to such developments with any certainty. Is there something about altering biology (e.g., through the artificial manipulation of DNA) that is inherently more threatening to people in some societies, including highly developed ones, than artificial manipulation of the structure of materials? This remains little more than speculation, however, and so far there is no clear evidence that such a shift is taking place.

For practicing risk communicators, the jury is still out on whether future events will spur a period of social amplification of risk for nanotechnology. Perhaps the right trigger simply remains to be activated. Or perhaps the era of biotechnology controversies was unique and has left us with a false expectation about technology-skeptical people. Where biotechnology risks seemed to have been socially amplified, nanotechnology's risks seem (by comparison) socially attenuated. While not as attention getting, the attenuation of risk (as opposed to its amplification) can be equally problematic for society as a whole, potentially leading to overconfidence in the corporate world or even lax regulation by governments whose action agendas may be set in large part by public insistence. Just because there are not huge numbers of anti-nanotechnology protestors making their voices heard today (and there are some, of course, but not many) does not mean there is no work for risk communication specialists.

Clearly, if all risk communicators needed to accomplish was to avoid people panicking, they would not have too much work to do in the case of nanotechnology just yet, even as their colleagues in social science struggle to figure out why. However, there are other sides to risk communication as a professional field. Finding ways to maintain and improve public trust, developing and providing new opportunities for public discussion, monitoring information about public opinion and media coverage, developing and maintaining contacts throughout a variety of stakeholder groups, making educational materials available that incorporate balanced discussions of risks and benefits, and keeping up with the science are work enough.

Proactive industrial organizations will benefit from investing in these activities. And some risk communicators, often those working for public interest or consumer advocacy organizations and other nongovernmental groups, are equally concerned that important risks not be ignored. Organizations like the Woodrow Wilson Center Project on Emerging Nanotechnologies have developed programs to encourage society and government to respond responsibly and appropriately to the risks of new technologies. These and other types of nonprofit organizations sometimes seek to balance our technological optimism with due caution, sometimes mobilizing publics and lobbying legislators to make sure risks are not ignored. Risk communicators work on these activities as well.

In the U.S. system of federal government, agencies such as the National Institute for Occupational Safety and Health, the National Institutes of Health, the Food and Drug Administration, the Department of Agriculture, and the Environmental Protection Agency, among others are all involved in the regulation and management of risks, including those associated with nanotechnology-based products and processes. Their work often involves risk communication, making information on risks (as well as benefits) available to citizens, workers, and employers as a means of reducing risk exposure. They serve a vital role by providing some of the most neutral and credible risk information available, information that is generally not associated with particular stakeholder groups. Equivalent organizations exist in other countries and (in the United States) at the state level; international organizations such as the Organization for Economic Development and Cooperation are also involved in this type of activity. Knowledge of risk communication informs the work of all of these agencies.

Public information officers of such agencies, as well as communicators who work on behalf of universities and other research agencies and organizations, also follow risk communication principles in their work, even though this may be only a part of their jobs. For example, a press release or other report of new research on nanomaterial toxicity needs to inform others about the research while avoiding either amplification or attenuation of the risk involved. Because research organizations, even not-for-profit ones, increasingly compete by seeking publicity for new research findings, providing material that gets attention but that is also as informative and neutral as possible can be a particular challenge. Academic research journals that publish primarily risk-related research (such as the journal *Risk Analysis*, published by the Society for Risk Analysis), as well as literally hundreds if not thousands of more general scientific publications around the world which may publish studies of this type on occasion, also seek to be recognized as leaders in the world of research. Their efforts to publicize the work that they publish also require knowledge of responsible risk communication practices. And the targets of all these efforts at information dissemination are often science journalists, who must understand the potential impact of information about risks.

In other words, just because the risks of nanotechnology have surprised observers by being less of a source of public concern than might have originally been anticipated, there is not necessarily any less work to be done by risk communicators. Some of these will be faced with trying to alert policy makers and broader publics to the existence of risks, while at the same time not acting as scaremongers. Meanwhile, somewhat reminiscent of the situation in the early years of biotechnology, governmental agencies have been struggling to determine whether their existing legal, regulatory, and procedural framework for managing risks is adequate for responding to these new developments in nanotechnology, and what new elements might be required. For example, nanotechnology involves such small-scale particles that regulations written to consider a certain minimum weight (that is, so many pounds or kilograms) or volume of material may leave important gaps. Certainly, reviewing the relevant regulatory frameworks for adequacy and applicability is necessary for public safety, although how much concern with regulation is too much remains a matter for debate involving competing philosophies of government, not just levels of risk. Even so, it seems likely that most stakeholders would agree that being as far ahead of the curve as possible in managing a novel set of risks is probably in everyone's best interests.

Here the "mad cow" history is especially instructive. When "mad cow" disease was first spreading in the United Kingdom and elsewhere, public officials (concerned with the economic impacts on the meat industry if concerned populations were to panic in the face of "mad cow") at first took great pains to reassure the public that this disease was under control and was not a threat to human health. When they turned out to be wrong on both counts (given that it is now generally accepted that a form of Creutzfeldt-Jakob disease in humans can come from consuming cattle afflicted with "mad cow"), public trust in government science was undoubtedly undermined. This means that the next time a similar situation arises, this same group of officials will face a much more difficult challenge in convincing various publics that they have the situation well in hand (Lanska 1998). Public officials (as well as corporate leaders) are well advised to be cautious and prudent in pronouncing any risk management challenge, including efforts to assess and control the risks of nanomaterials, completely under control. It is widely recognized that public trust is much more easily lost than regained.

Nanotechnology and "Cultural Resonance"

Despite the clear need for attention to responsible risk management, including risk communication, for nanotechnology, the question of why biotechnology has been the subject of so much public controversy while nanotechnology (thus far) has not remains largely unanswered. The phrase *cultural resonance,* sometimes used in the social movement literature in sociology, refers to the degree to which an argument or issue has salience or appeal, not just to individuals acting and reacting as individuals, but more generally across a cultural group (recalling here that both *collective behavior* and the *climate of opinion* for an issue involve group processes, not just individual psychology). Such groups constitute publics that (by definition) share many of their core values and beliefs. *Cultural resonance* refers to the dynamic whereby a particular interpretation or narrative element connects an issue or event with those shared values or beliefs, activates interest and concern within a group, and (on occasion) results in social mobilization to action.

Sometimes this resonance is linked to framing or problem definition processes that take place in mass media accounts of events, as well as in other forms of public and private discussion. Various social actors may actively promote particular frames or interpretations of a situation in the hopes of influencing members of target publics to interpret and react in a certain way (e.g., an automobile advertiser might seek to associate driving a certain kind of car with personal freedom and autonomy). But framing can also take place without a conscious strategic or persuasive intent. All stories (including media stories) have some kind of frame or structure that connects the elements of the story, and that may connect the story with deeply held values and beliefs.

Cultural reactions, positive or negative, can undoubtedly be induced to some extent by emphasizing some statements of fact and some interpretations over others. The degree of influence that such strategies might have on society as a whole is the subject of ongoing debate among scholars; in a free marketplace of ideas where citizens have access to many competing views, manipulating societal reactions is not easy, and the effects may be very short term. Yet sometimes, cultural resonance gives every appearance of being generated spontaneously, arising out of the culture rather than from conscious attempts to mobilize or persuade members of that culture. In this respect, nanotechnology simply does not seem to have this kind of built-in or inherent resonance to levels associated with previous technologies (such as biotechnology) that seemed to become controversial much more quickly. In the case of GM foods, the result was that a variety of advocacy groups became motivated to either protest or defend their further development and use. Setting aside the thorny issue of whether one group of technologies is inherently more dangerous than another, polarization was apparent very early in the GM foods controversy, while nanotechnology discussions seem to be following a different trajectory.

Several specific elements have been identified in our discussion up to this point that may have tended to amplify public reaction to biotechnology, including mistrust of corporate interests, less than full public transparency as biotechnology was first introduced, concern over regulatory adequacy for what appeared to be a completely novel set of technologies, and an apparent lack of obvious consumer benefit sufficient to outweigh even low-level consumer concerns. Because nanotechnology research in its early stages has been spurred by an initial investment by government research agencies as much as, or more than, by private corporations, relative levels of public trust in government versus private interests in research and development may also be a factor.

The term *cultural resonance* has been introduced here to suggest something deeper, more intricately intertwined with cultural values and norms (that is, beliefs about good and bad, acceptability and unacceptability). Trust, transparency, regulatory capacity, and consideration of benefits are all likely to be important contributors to how society reacts to a risk. However, cultural resonance is intended to mean something more, a gestalt reaction that occurs at a group level (and may lead to social amplification of a risk). It is not something we can entirely predict, at least not yet, and do not entirely understand, but it seems to be an important dynamic of our collective decision-making processes. In retrospect, it appears that biotechnology may have challenged our fundamental values and beliefs, and perhaps even our underlying sense of self, more directly than any form of material science will ever do. Issues of what people anticipate (and what they hope for) from technology more generally are also implicated here.

The distinction between life and nonlife—a distinction that is deeply engrained in many cultural traditions and reflected in familiar phrases such as "the sanctity of life"—was strongly challenged by biotechnology. It is currently being challenged further by developments in synthetic biology and synthetic genomics, techniques that promise to give us the power to actually create life from nonlife. Synthetic genomics researchers have been combining genes from completely unrelated organisms for a number of years, producing garden slugs and even pigs turned yellow by a jellyfish gene. This goes well beyond what "conventional" genetic modification techniques can generally do; the researcher has much more control over the processes involved because they can literally manufacture the necessary genetic material. These researchers have already created self-replicating life by artificially producing strands of DNA that are then inserted into a living cell, creating a new and previously nonexistent type of one-cell organism that is then capable of reproduction (Fox 2010).

If the life/nonlife distinction is a crucial element in why people seemed to react more strongly to biotechnology than to nanotechnology, then we might suppose that synthetic genomics would be the most controversial. Yet, so far, this has not proven to be the case. Perhaps this is because the organisms in question are generally quite simple, to date. Perhaps it is because the researchers involved have gone to considerable effort to generate broad discussion, including discussion of ethical dimensions, at the very early stages of this research, following good risk communication practice. Clearly, though, much remains to be learned about why some technologies and their associated risks prove controversial, while others do not.

PERSPECTIVE: PUBLIC ATTITUDES TOWARD NANOTECHNOLOGY

Dietram A. Scheufele

Public communications about nanotechnology such as news accounts may be mirroring issue cycles for previous technologies, including agricultural biotechnology. In particular, early coverage of nanotechnology was dominated by a general optimism about the scientific potential and economic impacts of this new technology (Dudo et al. 2011; Friedman and Egolf 2005, 2007). This is in part related to the fact that a sizable proportion of nanotechnology news coverage, at least in newspapers, continues to be provided by a handful of science journalists and business writers (Dudo et al. 2011).

WHERE PUBLIC ATTITUDES COME FROM

The overall positive framing of nanotechnology in news outlets is also linked to support for more research and funding among the general public (Cobb and Macoubrie 2004; Scheufele and Lewenstein 2005). This connection between media coverage and support for nanotechnology, however, does not follow traditional knowledge deficit models (for an overview, see Brossard et al. 2005). Instead, most research on public attitudes toward nanotechnology does not show an impact of media coverage on lay audiences' understanding of the technology, which, according to knowledge deficit models, would produce more positive attitudes toward the technology. Most recent research has found that the driving factors behind public attitudes are various forms of heuristics or cognitive shortcuts that audiences use to make sense of the technology, even in the absence of information (Scheufele 2006).

One of these heuristics is media frames. Frames are ways of presenting an issue that will produce particular outcomes among audiences (Scheufele 1999). Framing is often traced back to Nobel Prize winning work in experimental psychology that examined how embedding information in particular contexts can shape people's interpretation of that information (Kahneman 2003). When applied to mass media, framing theory suggests that even small terminological tweaks in terminology ("nano as the next plastic" versus "nano as the next asbestos") can activate different cognitive frameworks among audiences and shift the interpretation of the technology overall (Scheufele and Tewksbury 2007).

How people think about nanotechnology or what cognitive schemas they use to make sense of nano-related information, however, also depends on the specific aspect of nanotechnology that is being discussed. Nanotechnology has often been described as an enabling technology. The *nano* label, in other words, simply describes work or observations at a particular size scale. As a result, nanotechnology research bridges a diverse set of research fields and application areas.

Attitudes depend heavily on the specific area that is being discussed at any given moment. Recent national survey data, for instance, show that people's likelihood of translating their perceptions of risks associated with nanotechnology to specific attitudes about the technology depends to a significant degree on the specific application area they think about when forming those risk-attitude judgments (Cacciatore et al. forthcoming).

WHAT TYPES OF ATTITUDES AND COGNITIONS MATTER MOST?

Virtually all national surveys tapping public opinion on nanotechnology have measured some form of knowledge or at least perceived knowledge. This distinction, unfortunately, is lost to many commentators who often conclude in a simplistic fashion that the public knows little about nanotechnology, even if a study relies on self-reported awareness of the technology.

Empirically, however, self-reported perceptions and objective assessments of knowledge are clearly distinct. The former taps people's perceptions of what they *think* they know. A number of researchers, for example, have tapped perceived levels of awareness about nanotechnology (Peter D. Hart Research Associates 2006, 2007; Scheufele and Lewenstein 2005) and have shown little change over time. Significantly fewer studies have actually tracked objective quiz-type measures of what the public knows about nanotechnology (Lee and Scheufele 2006). Typically, such measures include a battery of true-and-false-type knowledge questions about nanotechnology.

From the data available, we can see two trends: First, levels of knowledge about nanotechnology across the general population have remained fairly static in the last few years (Scheufele et al. 2009, see Online Appendix); second, we see a widening gap among education groups, with highly educated respondents showing increased learning over time, and less-educated respondents falling behind in terms of how much they know about nanotechnology (Corley and Scheufele 2010).

Aside from cognitive variables, research on public attitudes toward nanotechnology has also explored overall attitudes toward nanotechnology. Most of this research has shown that people's views on nanotechnology are generally positive (Cobb and Macoubrie 2004). Respondents who self-identify as being more aware of nanotechnology tend to show higher levels of overall support than respondents who are less aware of nanotechnology (Scheufele and Lewenstein 2005).

A second attitudinal variable that has dominated research on public reactions to nanotechnology is people's judgment about the relative risks and benefits of nanotechnology. Across studies, patterns of results suggest that those who perceive greater benefits for nanotechnology outnumber those who perceive greater risks by 3 to 1 (Satterfield et al. 2009). Unfortunately, previous research has relied mostly on a single item to tap these relative assessments of risks and benefits among the general public: "Do the risks associated with nanotechnology outweigh the benefits; do the benefits outweigh the risks; or are the risks and benefits approximately the same?"

A number of researchers have raised serious concerns about these measures and their potential to provide invalid assessments of risk perceptions among the general public. At a conceptual level, these criticisms have focused on at least two areas. First, responses may be biased based on response order effects. Asking respondents first whether "the benefits outweigh the risks," followed by response options for "the risks outweighing the benefits," or "risks and benefits being about equal," for instance, is a much different question than one that offers the "risks outweighing the benefits" as the first response option. Second, single-item measures force respondents to make subjective summative judgments about the relative importance of several risks and benefits. Such judgments, unfortunately, are often skewed, given people's tendency to remember unfavorable information about a topic better than favorable information.

Most recently, however, Binder and colleagues (forthcoming) quantified the potential response biases introduced by single-item measure of risk and benefits perceptions. Specifically, their comparisons of results from two surveys and across different measures of risk/benefits perceptions suggest that single-item measures of risk and benefits perceptions may be slanting answers toward higher risk perceptions. People perceived more benefits than risks when given the opportunity to evaluate these attributes separately, as opposed to being asked to make a quick summary judgment in a single item. Interestingly, this pattern holds for both issues tested in the study—biofuels and nanotechnology.

EXPERT OPINIONS VERSUS PUBLIC OPINION

A growing body of research is also beginning to compare attitudes among members of the lay public to expert surveys. Most systematic surveys among U.S. nanoscientists suggest that they are more optimistic than the general population about the potential benefits of nanotechnology, and, in most areas, less pessimistic about its potential risks (Besley et al. 2008; Scheufele et al. 2007). Comparisons of answers to identically worded questions in surveys among the leading nanoscientists in the United States and a representative sample of the U.S. population, however, showed that there were two areas in which nanoscientists showed higher levels of concern about potential risks of nanotechnology than the general public: human health and environmental pollution (Scheufele et al. 2007).

Previous research has also examined to what degree experts' opinions on nanotechnology are driven by different factors than opinions among the lay public (Ho et al. 2010; Priest et al. 2010). Not surprisingly, much of this research shows that attitudes among nanotechnology experts are strongly correlated to their scientific judgments about potential risks and benefits. What is interesting, however, is the fact that experts' stances on stricter regulations for nanotechnology are, at least in part, driven by their political viewpoints, even after their judgments on batteries of questions about objective risks and benefits are taken into account. More conservative scientists tend to also be more opposed to stricter regulations, whereas liberal-leaning scientists tend to support them (Corley et al. 2009).

REMAINING CHALLENGES

Two challenges are emerging as public attitudes toward nanotechnology develop along with the technology. The first challenge relates to a long-standing problem surrounding the development of technical innovations in modern societies: knowledge gaps. Knowledge gaps do not simply refer to different levels of understanding about a technology across social groups. Instead, the concept goes back to work by Tichenor and colleagues (1970) who showed that learning effects from informational campaigns were significantly higher among respondents with high socioeconomic status (SES) than respondents with lower levels of SES.

For nanotechnology, we see similar patterns emerge. Recent analyses of nationally representative trend data (Corley and Scheufele 2010) show widening gaps for knowledge about nanotechnology between the most and least educated groups in the United States. In other words, as the technology evolves and has an impact on more and more areas of our daily lives, highly educated respondents become more familiar with nanotechnology and its applications, but less educated groups fall behind and are potentially becoming less and less informed about nanotechnology as societal debates focus on an increasingly complex set of ethical, legal, and social challenges (Khushf 2006).

This has tremendous implications for many outreach efforts, such as nano cafes or museum exhibits, that traditionally target a more interested and informed segment of the population and may be less effective as channels for reaching disadvantaged or harder-to-reach audiences. But there is a silver lining. A closer look at the media use patterns among different SES groups shows that online sources of information about nanotechnology can help overcome knowledge deficits for low SES respondents (Corley and Scheufele 2010), and future research will have to explore how to better utilize online communication channels to more systematically target hard-to-reach audiences.

A second challenge for researchers studying public attitudes toward nanotechnology is the role that personal values play in helping people make sense of new information about emerging technologies. Previous research has shown how religious views (Brossard et al. 2009), cultural predispositions (Kahan et al. 2008, 2009), and views about scientific authority (Brossard and Nisbet 2007; Lee and Scheufele 2006) shape how people translate (mass mediated) information into attitudes toward nanotechnology. In other words, values and predispositions can serve as *perceptual filters* (Brossard et al. 2009) that shape information processing, and the same piece of information will be interpreted very differently by different audiences, depending on their preexisting values and predispositions.

This role of values as perceptual filters is particularly important given recent comparisons among the United States and various European countries. These comparisons showed significant variation in religious views across countries and also a significant relationship between those views and attitudes toward nanotechnology (Scheufele et al. 2009). As regulators in the United States work with their counterparts in other countries in order to harmonize regulatory frameworks for nanotechnology, understanding the value landscape in each country will be absolutely critical for evaluating the viability of regulatory choices and restrictions. Future research will have to much more systematically examine public attitudes toward nanotechnology and its applications in an international context.

REFERENCES

Besley, J., V. Kramer, and S. Priest. 2008. Expert opinion on nanotechnology: risk, benefits, and regulation. *Journal of Nanoparticle Research* 10(4): 549–558.

Binder, A.R., M.A. Cacciatore, D.A. Scheufele, B.R. Shaw, and E.A. Corley. Forthcoming. Measuring risk/benefit perceptions of emerging technologies and their potential impact on communication of public opinion toward science. *Public Understanding of Science*. doi: 10.1177/0963662510390159 (available online at pus.sagepub.com).

Brossard, D., B.V. Lewenstein, and R. Bonney. 2005. Scientific knowledge and attitude change: the impact of a citizen science project. *International Journal of Science Education* 27(9): 1099–1121.

Brossard, D., and M.C. Nisbet. 2007. Deference to scientific authority among a low information public: understanding U.S. opinion on agricultural biotechnology. *International Journal of Public Opinion Research* 19(1): 24–52. doi: 10.1093/ijpor/edl003.

Brossard, D., D.A. Scheufele, E. Kim, and B.V. Lewenstein. 2009. Religiosity as a perceptual filter: examining processes of opinion formation about nanotechnology. *Public Understanding of Science* 18(5): 546–558. doi: 10.1177/0963662507087304.

Cacciatore, M.A., D.A. Scheufele, and E.A. Corley. Forthcoming. From enabling technology to applications: the evolution of risk perceptions about nanotechnology. *Public Understanding of Science*. doi: 10.1177/0963662509347815.

Cobb, M.D., and J. Macoubrie. 2004. Public perceptions about nanotechnology: risks, benefits and trust. *Journal of Nanoparticle Research* 6(4): 395–405.

Corley, E.A., and D.A. Scheufele. 2010. Outreach gone wrong? When we talk nano to the public, we are leaving behind key audiences. *The Scientist* 24(1): 22.

Corley, E.A., D.A. Scheufele, and Q. Hu. 2009. Of risks and regulations: how leading U.S. nanoscientists form policy stances about nanotechnology. *Journal of Nanoparticle Research* 11(7): 1573–1585. doi: 10.1007/s11051-009-9671-5.

Dudo, A.D., D.-H. Choi, and D.A. Scheufele. 2011. Food nanotechnology in the news. Coverage patterns and thematic emphases during the last decade. *Appetite* 56(1): 78–89. doi: 10.1016/j.appet.2010.11.143.

Dudo, A.D., S. Dunwoody, and D.A. Scheufele. 2011. The emergence of nano news: tracking thematic trends and changes in U.S. newspaper coverage of nanotechnology. *Journalism & Mass Communication Quarterly* 88(1): 55–75.

Friedman, S.M., and B.P. Egolf. 2005. Nanotechnology: risks and the media. *IEEE Technology and Society Magazine* 24: 5–11.

Friedman, S.M., and B.P. Egolf. 2007. Changing patterns of mass media coverage of nanotechnology risks. Paper presented at the Project on Emerging Nanotechnologies, Woodrow Wilson Center for International Scholars (December 18).

Ho, S.S., D.A. Scheufele, and E.A. Corley. 2010. Making sense of policy choices: understanding the roles of value predispositions, mass media, and cognitive processing in public attitudes toward nanotechnology. *Journal of Nanoparticle Research* 12(8): 2703–2715. doi: 10.1007/s11051-010-0038-8.

Kahan, D.M., D. Braman, P. Slovic, J. Gastil, and G. Cohen. 2009. Cultural cognition of the risks and benefits of nanotechnology. *Nature Nanotechnology* 4(2): 87–90. doi: 10.1038/nnano.2008.341.

Kahan, D.M., P. Slovic, D. Braman, J. Gastil, G. Cohen, and D. Kysar. 2008. Biased assimilation, polarization, and cultural credibility: an experimental study of nanotechnology risk perceptions. *Project on Emerging Nanotechnologies Research Brief No. 3*.

Kahneman, D. 2003. Maps of bounded rationality: a perspective on intuitive judgment and choice. In *Les Prix Nobel: The Nobel Prizes 2002*. Ed. T. Frängsmyr. Stockholm, Sweden: Nobel Foundation, 449–489.

Khushf, G. 2006. An ethic for enhancing human performance through integrative technologies. In *Managing Nano-Bio-Info-Cogno Innovations: Converging Technologies in Society*. Eds. W.S. Bainbridge and M.C. Roco. Dordrecht, The Netherlands: Springer, 255–278.

Lee, C.J., and D.A. Scheufele. 2006. The influence of knowledge and deference toward scientific authority: a media effects model for public attitudes toward nanotechnology. *Journalism and Mass Communication Quarterly* 83(4): 819–834.

Peter D. Hart Research Associates. 2006. "Public awareness of nano grows—majority remain unaware." The Woodrow Wilson International Center for Scholars Project on Emerging Nanotechnologies. Retrieved October 3, 2006, from www.nanotechproject.org/78/public-awareness-of-nano-grows-but-majority-unaware.

Peter D. Hart Research Associates. 2007. Poll reeals public awareness of nanotech stuck at low level. Retrieved October 10, 2007, from http://www.nanotechproject.org/news/archive/poll_reveals_public_awareness_nanotech.

Priest, S., T. Greenhalgh, and V. Kramer. 2010. Risk perceptions starting to shift? U.S. citizens are forming opinions about nanotechnology. *Journal of Nanoparticle Research* 12(1): 11–20. doi: 10.1007/s11051-009-9789-5.

Satterfield, T., M. Kandlikar, C.E.H. Beaudrie, J. Conti, and B. Herr Harthorn. 2009. Anticipating the perceived risk of nanotechnologies. *Nature Nanotechnology* 4(11): 752–758. doi: 10.1038/nnano.2009.265.

Scheufele, D.A. 1999. Framing as a theory of media effects. *Journal of Communication* 49(1): 103–122.
Scheufele, D.A. 2006. Messages and heuristics: how audiences form attitudes about emerging technologies. In *Engaging Science: Thoughts, Deeds, Analysis and Action*. Ed. J. Turney. London: The Wellcome Trust, 20–25.
Scheufele, D.A., E.A. Corley, S. Dunwoody, T.-J. Shih, E. Hillback, and D.H. Guston. 2007. Scientists worry about some risks more than the public. *Nature Nanotechnology* 2(12): 732–734.
Scheufele, D.A., E.A. Corley, T.-J. Shih, K.E. Dalrymple, and S.S. Ho. 2009. Religious beliefs and public attitudes to nanotechnology in Europe and the U.S. *Nature Nanotechnology* 4(2): 91–94. doi: 10.1038/NNANO.2008.361
Scheufele, D.A., and B.V. Lewenstein. 2005. The public and nanotechnology: how citizens make sense of emerging technologies. *Journal of Nanoparticle Research* 7(6): 659–667.
Scheufele, D.A., and D. Tewksbury. 2007. Framing, agenda setting, and priming: the evolution of three media effects models. *Journal of Communication* 57(1): 9–20. doi: 10.1111/j.1460-2466.2006.00326.x.
Tichenor, P.J., G.A. Donohue, and C.N. Olien. 1970. Mass media flow and differential growth in knowledge. *Public Opinion Quarterly* 34(2): 159–170.

References

Bainbridge, W.S. 2002. Public attitudes toward nanotechnology. *Journal of Nanoparticle Research* 4(6): 561–570.
Bergstein, B. 2008, May 20. "Study: carbon nanotubes mimic asbestos in mouse tests." *USAToday*. www.usatoday.com/tech/news/nano/2008-05-20-carbon-nanotubes-asbestos_N.htm.
Bishop, G.F. 2005. *The Illusion of Public Opinion: Fact and Artifact in American Public Opinion Polls*. Oxford: Rowman and Littlefield.
Currall, S.C. 2009. Nanotechnology and society: new insights into public perceptions. *Nature Nanotechnology* 4(2): 79–80.
Fox, S. 2010, May 21. "J. Craig Venter Institute creates first synthetic life form." *Christian Science Monitor*. www.csmonitor.com/Science/2010/0521/J.-Craig-Venter-Institute-creates-first-synthetic-life-form.
Fujita, Y., H. Yokoyama, and S. Abe. 2006. Perception of nanotechnology among the general public in Japan—of the NRI nanotechnology and society survey project. *Asia Pacific Nanotech Weekly* 4: 1–2. www.nanoworld.jp/apnw/articles/library4/pdf/4-6.pdf.
Gaskell, G., T.T. Eyck, J. Jackson, and G. Veltri. 2005. Imagining nanotechnology: cultural support for technological innovation in Europe and the United States. *Public Understanding of Science* 14(1): 81.

Kahan, D.M., D. Braman, P. Slovic, J. Gastil, and G. Cohen. 2009. Cultural cognition of the risks and benefits of nanotechnology. *Nature Nanotechnology* 4(2): 87–90.

Kasperson, J.X., R.E. Kasperson, N. Pidgeon, and P. Slovic. 2003. The social amplification of risk: Assessing fifteen years of research and theory. 13–46. In *The Social Amplification of Risk,* Eds. N. Pidgeon, R. Kasperson, and P. Slovic. London: Cambridge University Press, 13–46.

Lanska, D.J. 1998. The mad cow problem in the U.K.: risk perceptions, risk management, and health policy development. *Journal of Public Health Policy* 19(2): 160–183.

Miller, J.D. 1986. Reaching the attentive and interested publics for science. In *Scientists and Journalists,* Eds. S. Friedman, S. Dunwoody, and C. Rogers. Washington, DC: American Association for the Advancement of Science, 55–69.

Pidgeon, N., B.H. Harthorn, K. Bryant, and T. Rogers-Hayden. 2008. Deliberating the risks of nanotechnologies for energy and health applications in the United States and United Kingdom. *Nature Nanotechnology* 4(2): 95–98.

Poland, C.A., R. Duffin, I. Kinloch, A. Maynard, W.A.H. Wallace, A. Seaton, V. Stone, S. Brown, W. MacNee, and K. Donaldson. 2008. Carbon nanotubes introduced into the abdominal cavity of mice show asbestos-like pathogenicity in a pilot study. *Nature Nanotechnology* 3(7): 423–428.

Priest, S.H. 2001. *A Grain of Truth: The Media, the Public, and Biotechnology*. Lanham, MD: Rowman and Littlefield.

Priest, S., and A. Gillespie. 2000. Seeds of discontent: scientific opinion, the mass media and public perceptions of agricultural biotechnology. *Science and Engineering Ethics* 6(4): 529–539.

Priest, S., T. Greenhalgh, and V. Kramer. 2010. Risk perceptions starting to shift? U.S. citizens are forming opinions about nanotechnology. *Journal of Nanoparticle Research* 12(1): 11–20.

Priest, S., T.B. Lane, T. Greenhalgh, L.J. Hand, and V. Kramer. In press. Envisioning emerging nanotechnologies: a three-year panel study of South Carolina citizens.

Sale, K. 1999. The achievements of "General Ludd": a brief history of the Luddites. *Ecologist* 29(5): 310–313.

Satterfield, T., M. Kandlikar, C.E.H. Beaudrie, J. Conti, and B.H. Harthorn. 2009. Anticipating the perceived risk of nanotechnologies. *Nature Nanotechnology* 4(11): 752–758. doi:10.1038/NNANO.2009.265.

5

What Do People Want from Technology?

Discussions about encouraging public engagement and public consultation with respect to decisions about science and technology policy are taking place all over the world, especially in Europe, Canada, and (increasingly) the United States. Public engagement activities are seen as a way of increasing public understanding of science. These take place in a variety of formats that are meant to be more interactive than an ordinary lecture, ranging from science cafes, meaning organized activities in which scientists and citizens can have informal chats and interpersonal discussions, consensus conferences and citizen forums, where nonscientists are invited to discuss their views of scientific developments and (in some cases) recommend policies, or demonstrations and exhibits at science museums that feature opportunities for interaction between experts and nonexperts. All of these activities routinely assume that "ordinary" citizens (that is, members of various nonexpert publics) have something to contribute to the decision-making process. The idea of authentic two-way risk communication, in principle, involves the same assumption. The recognition that judgments about the acceptability of risks is a matter of opinion, driven by values and not just scientific facts, is strong support for consulting citizens about where technology should be going.

However, organizing such discussions has sometimes been primarily reactive, arising only once negative public opinion emerges (or is suspected to be emerging). Even though the argument that broader publics should be consulted earlier in the process of deciding public policy with respect to technology is becoming more common, the science community sometimes has unrealistic hopes about what this will accomplish. Early in the history of an emerging technology, these types of activities, which are sometimes referred to as "upstream" public engagement in technology policy, seek to get members of various publics involved before the climate of public opinion has been fully formed. Crucially, though, public engagement activities should not be viewed as a means to head off negative public reaction. Organizing engagement activities should not be undertaken with the goal of deflecting opposition. If the purpose is really to allow nonexperts to reflect on science and express their views, the end result could be anywhere along a continuum from negative to positive. Often, members of the science and science policy communities have seized on the idea of *public engagement* as a means to encourage support for science.

This is a risky proposition and not a good argument for engaging nonscientific "publics."

Yet, is it also possible for public engagement to occur too far upstream? For nanotechnology, with low levels of awareness and familiarity and very little apparent *cultural resonance* that might stimulate public interest, modest awareness of risks has not yet produced much in the way of public concern. Issues of who will participate, if meaningful opportunities for public discussion are made available, are important. If public opinion surveys (which can be seen as a weak, arguably the very weakest, form of engaging and consulting the public) have difficulty obtaining random samples, asking people to give up anywhere from a few minutes to several hours to, in some models, even several days of free time to discuss future technology, engagement and consultation activities will have many times more difficulty in attracting participants from all walks of life—that is, from multiple broader publics—who are willing to participate.

If this discussion would appear to have moved beyond consideration of communicating risks, that is probably a valid observation. Perceptions of risks, as well as judgments about which risks are acceptable, are inextricably intertwined with social and cultural values, beliefs, and norms. Further, when technologies bring tangible benefits, this may motivate people to accept associated risks (as for consumer technologies ranging from cell phones to automobiles.) The task of risk communication is rather more complex than simply "translating" expert evaluations into a form that nonexperts can understand. An important and immediately relevant theme of the science-technology-and-society literature is that technologies are generally driven by social and cultural priorities, not the other way around. We create and adopt the technologies that we believe we need—those that seem to serve our individual collective interests.

Technological Literacy and Democracy

Democratic societies can be conceptualized in various ways. One common distinction is between the notion of *elite pluralism,* in which a variety of groups representing various stakeholders with competing interests are conceptualized as competing for power and influence, versus a more "pure" form of democracy in which individual citizens debate and discuss issues, discussions that then directly or indirectly inform the policymaking process. In an idealized version of the elite pluralism model, the various competing interests would tend to balance one another out, so that no one group has excessive influence; however, this is a competition among the powerful. This might be seen as one variant of the *marketplace of ideas* model described earlier in this book, but it is a variant that does not especially recognize (or criticize) the relationship between influence and political power. Thus, the chief criticism of this model is that only a minority of "players" can appear (effectively) on this stage—those with disproportionate political and economic power. Although this description might accurately describe how modern governance actually takes place in most democracies, it is a vision in which ordinary citizens who are not associated with (that is, who are not represented by) powerful organized groups tend to have little say. Absent effective labor unions, for example, workers are underrepresented and have little power.

A competing model that sees debate and discussion among ordinary citizens as at the center of democracy is today often associated with the European philosopher Jürgen Habermas (see Chapter 2), who popularized the idea of the *public sphere* as a way of describing the sites where those discussions take place. These public sphere discussions can take place anywhere people tend to gather and discuss events of the day: in cafes or pubs or hair salons or living rooms, on street corners, and so on. Another way of thinking about a more direct democracy, as suggested by the concept of *deliberative democracy*, is a perspective that imagines a more participatory approach, but one likely involving consciously structured events in which people are specifically encouraged to express and discuss their views. For science and technology, the current movement toward providing new opportunities for deliberation and discussion through public engagement fits here. Rather than seeing politically relevant conversations as arising in more or less naturally occurring forums (say, on street corners), realizing the idea of deliberative democracy may require active intervention, particularly for science, to which everyone may not be attracted (but by means of which everyone is affected).

In public participation exercises inspired by the concept of deliberative democracy, citizens may be invited to discuss their views in a slightly more formal way, with their discussions sometimes leading to an identifiable consensus position under some circumstances (but not all). The active and deliberate organization of political debate among everyday citizens is often associated with Scandinavian political traditions but has spread around the world, especially with respect to science and technology policy discussions as recognition grows that involving broader publics in these discussions is important.

Similarly, for public opinion, there have been two competing views that have conceptualized its role and foundation quite differently, as captured in a well-known disagreement between former journalist and early public opinion researcher Walter Lippman and educational philosopher John Dewey in the first half of the twentieth century. Lippman felt that ordinary people were really not capable of understanding the full range of complex issues facing modern societies at any given moment, and were therefore readily subject to manipulation by media accounts that represented these issues in one way or another, always giving, at most, no more than a partial view of complex issues. Dewey believed that, on the contrary, with meaningful education all citizens could, at least potentially, become informed and have meaningful, reasoned opinions on the full range of issues of the day. This arguably less cynical view was the foundation of an educational philosophy that emphasized imparting the skills of citizenship, not just the skills needed for a job.

The concept of pure or direct democracy is often traced back to ancient Greece, but the Greeks owned slaves, and only a relatively few Greek individuals actually qualified as "citizens" in that system of democracy, which largely excluded women as well as slaves. One of the criticisms of Habermas' "public sphere" democracy is that it is (similarly) based on his thinking about eighteenth-century European society, a society in which only a minority of people were in an economic position to have the twin luxuries of education and free time available for political discussion. Anyone who has observed political debate in a modern country like Italy, for example, where the tradition of street-corner debate is still alive and well, recognizes that impassioned discussion of politics need not be limited to the elite "drawing room" discussions of an emerging bourgeois class to which Habermas sometimes referred. The extent to which Habermas' vision is entirely democratic is an ongoing debate in itself (Habermas 1985).

However, scientific and technological literacy remain limited. How can average people participate in a meaningful way in discussions of the future of nanotechnology or (say) of something like synthetic biology? These questions of the appropriate role of the average citizen and of who actually belongs in a political debate about complicated issues remains in the background of many contemporary policy discussions, especially those about emerging technology and cutting-edge science. On the one hand, the full implications of the scientific foundations of technological developments in areas such as nanotechnology are fully understood by only a minority, at best; on the other hand, these affect all of us, and it is reasonable to argue that all citizens in a technology-based democracy can (and should) have sensible and rational expectations for what they expect from technology and how it should, in general, be managed. These concerns surrounded the management, in the city of Cambridge, of the earliest experiments in genetic engineering and remain relevant today (Wright 1994).

In general, most modern democracies are organized as representative democracies, certainly not as direct democracies, if only for practical reasons, especially when we are thinking beyond the most local level. Taking the United States as a convenient example, in the early days of the Western frontier, it was hardly possible for most ordinary citizens to journey to

Washington, DC, on the far-away East Coast, or even to a nearby state capital, to participate in political discussions. Newspapers brought news of political events and discussions slowly, only well after the fact, and were originally too costly to be available to anyone but economic elites. Electing representatives to speak on their behalf was the only possible way everyday citizens could get their voices heard on many matters, particularly at the federal level; this was not a particularly philosophical decision at the time, but a very practical one about how to get the work of democratic government done. Representative democracy did not develop as a philosophical ideal, but as a practical necessity.

More recently, first mass-market newspapers, then broadcast news, and then (most recently) the Internet regularly transmit information about events, including political events, instantly, and on a 24-hour, 7-day cycle, providing ready access to information and opinion on just about everything, and also providing potential new forums for citizen debate and discussion across the political spectrum, given "new media's" capacity for interactivity. Even so, it is not entirely practical to propose the practice of direct democracy in today's larger societies, even with the help of modern communication technology. How could millions or in some cases billions of individuals simultaneously participate in national discussions, regardless of how effectively the new communication technologies serve us? And no matter how intelligent, dedicated, or well educated, can all of us keep up with all of the issues of the day, such that we are always ready and able to participate in discussions about how to respond to them? Profound challenges to the practice of pure democracy remain.

The Challenges of Risk Society

It is somewhat remarkable that many modern democracies are as stable and prosperous as they are, given the constraints and challenges of collective decision making involving so many people of diverse backgrounds. Our imperfect tools for both popular and political discussion are the best we have, yet they are not ideal. Issues related to the management of complex technology based on recently emerging science, management carried out by an enormously complicated web of government agencies, represent especially challenging areas. Yet, if technology embodies human values, broad democratic participation in these decisions would appear essential. In the United States, federal, state, and local agencies and governments share power over these decisions (ranging from the management of federally owned lands to the regulation of nanotechnology); in other large bureaucracies such as those of the European Union or the governments of India or

China, equally complex constellations of national (even supranational) and provincial or state jurisdictions similarly both compete and cooperate in creating and managing public policy. And this is to mention only a handful of the geographically larger political entities in the modern world.

The term *technocracy* was invented to describe the bureaucratic structure of a modern, complex (and yet at least nominally democratic) society in which individual regulators and managers must be highly specialized governance experts, often trained as scientists or social scientists. The term seems particularly applicable to the most highly developed democracies, those with economies and lifestyles that are intimately intertwined with the state of technology and science, and with increasingly ethnically diverse populations that bring significantly different values and perspectives to the table. Highly trained regulators make many decisions on behalf of us all.

And there are certainly strains and stresses that result. For example, technologies bring risks as well as benefits. What does it mean to distribute those risks and those benefits in a just and democratic way? How is this accomplished? Are we taking the right steps to manage the environmental and health risks of technology, including nanotechnology? Toxicological research, along with relevant regulatory capacity, has been hard-pressed to keep up with developments in nanotechnology, as in other areas of other technological developments. The U.S. Gulf Coast oil disaster (April 20, 2010) illustrates that we do not always have technology as much under control as we think we do. And the prospect of global climate change shows just how difficult it is for even our most powerful, most scientifically sophisticated, most technocratically managed societies to switch gears in time to prevent looming worldwide disaster. Broad-based citizen participation is also essential.

The citizens of today's technology-based societies need a new form of literacy, referred to as technological literacy (distinct from science literacy). As a 2002 report from the U.S. National Academy of Engineering notes, even today's engineers "may not have the training or experience necessary to think about the social, political, and ethical implications of their work *and so may not be technologically literate*" (Pearson and Young 2002, 22, italics added). In other words, understanding the societal implications of technology is considered a core part of technological literacy, and, as this same report also goes on to argue, technological literacy is essential for people in a modern democracy to function effectively, both as consumers and as citizens.

Technologies shape our present and our future. The choices we make today about what kinds of research and development are worthy of investment, how to manage technology's risks and distribute its benefits, what policies should be put into place to encourage the adoption of particular technologies (e.g., those that hold the promise of delivering cleaner energy), what restrictions should be placed on others (e.g., those judged too risky, too polluting, or otherwise problematic for society's broader interests), and what products we choose to purchase as individual consumers go a long way toward determining what kind of society we will live in tomorrow. However, only rarely do we take time out to consider what kind of society we would want that to be, and how technology might help (or impede) those goals. An ideal program of public engagement might encourage citizens to consider these long-term goals, but such efforts to date are only a proverbial drop in the bucket.

Extensive interviews with a broad range of individuals (students, scientists, and, above all, ordinary citizens) designed to elicit their initial impressions about nanotechnology confirm that we often look toward technology to improve our quality of life, and yet we also routinely anticipate that technology will not come without risks (Priest et al. in press). These recognitions are key ingredients of technological literacy. People apply these generalized expectations—what we might think of as a palate or "template" of technology-related concerns and expectations—to the evaluation of each new group of technologies, such as nanotechnology, even in cases where individual citizens have only a very general idea of what those new technologies might consist of. In other words, optimistically, citizens may have higher levels of technological literacy than we might be tempted to assume.

About two-thirds of the 76 ordinary citizens who initially participated in our 2007 interview study[1] were at least somewhat familiar with the term *nanotechnology*, and the idea struck a roughly equal proportion of them as generally positive, but nearly three-fourths were unfamiliar with any of the hundreds of nanotechnology-based products that were, even then, on the market. Their positive expectations can hardly be attributed to specific previous experiences with nanotechnology, but they reflect a set of initial expectations in which benefits were generally expected to outweigh risks, but issues were raised about both. Yet the fact that our nonexpert interviewees were completely unaware of the many existing products using nanotechnology suggests low awareness, consistent with our argument that nanotechnology has (for whatever reasons) been the subject of attenuated, rather than amplified, concerns that to date have reached only a narrow audience.

These *citizen panel* interviews about nanotechnology certainly suggested a partial victory in regard to technological literacy, in that individual participants generally offered thoughtful and informed perspectives on the likelihood that the unfamiliar new world of nanotechnology would provide both great benefits and noticeable risks. Their palate or what we

[1] Conducted in South Carolina among members of a wide range of urban, suburban, and rural community-based groups across socioeconomic lines; see Priest et al. (in press) for details.

called their *template* of hopes and fears suggest they have a reasonable and rational notion of what to expect from any new set of technologies, an important ingredient for technological literacy. But their low awareness of both the technology and its existing level of deployment in society suggest that the opportunities for public sphere discussion have been limited, as well as that our news and information networks have not furnished them with much in the way of raw material for such discussion, absent (ironically) solid evidence of risk.

Close to 70% of our interviewees, when engaged in in-depth discussions about their expectations for nanotechnology, identified some form of ethical concerns, contrary to the initial focus group results discussed in an earlier chapter. Some of our interviewees were concerned, for example, about the moral views of scientists, about science being out of control or getting misused, or about the risks affecting the poor and uneducated in disproportionate numbers. It seemed as though our participants tended to project both their hopes and their fears for technology onto their expectations for nanotechnology's future societal implications. On the one hand, our typical panel member was most likely to imagine medical benefits and to see these as the most important likely benefits of nanotechnology. On the other, risks of side effects or other negative consequences were often envisioned, alongside these possible benefits. Many of these participants had not thought much about nanotechnology before being asked to participate in our study (and even so, our study may well have been weighted toward those with more interest than average, who were thus more likely to agree to participate in an extended research project).

Our experiences thus confirm that ordinary citizens who are not necessarily very knowledgeable about a new set of technologies nevertheless have specific expectations about technology in general, including its societal implications, and that they apply these expectations proactively to a novel situation. Although cognitive and educational psychologists might say that these participants appeared to have *schemas* or categories representing technology in general, we prefer the term *template* to suggest that these categories are like forms waiting to be filled out, forms with many blanks. Yet the levels of knowledge and familiarity of our panelists, specifically for nanotechnology, were limited.

Technology is not introduced in a vacuum; the minds of citizens and consumers contemplating the adoption of new technologies are not completely blank slates. Yet, consistent with the results from the public opinion polls discussed in Chapter 3, this certainly does not mean that their expectations are always negative. Our technology-based lifestyle rests on a foundation of generally positive attitudes toward technology, and while this circumstance is sometimes belied by heated disagreements about adopting particular technologies, we might better interpret such cases as the exceptions rather than the rules of how we tend to react. The question of exactly how, in a democratic society, the perspectives of these citizens and consumers ought to be taken into account in making decisions about investing in, adopting, and managing particular technologies remains largely unanswered.

Nanotechnology, Risk, and Society

Upstream engagement has been encouraged by the Royal Society in the United Kingdom (Rogers-Hayden and Pidgeon 2007), the European Commission (Nordmann 2004), and the U.S. Congress (Roco 2003; 15 U.S.C. 7501 et seq.), among many other powerful groups of actors. Motivations for this may be mixed, with those in favor of improving democratic practice becoming politics' proverbial "odd bedfellows" with those who wish to avoid a public stalemate on technology policy for a variety of other reasons. Today's policy makers clearly recognize that societal values are important to public reactions to technology, and they also acknowledge (at least implicitly) that it is important to their own interests to consider this relationship in their decision making. The U.S. Congress did so in its passage of the 2003 21st Century Nanotechnology Research and Development Act (15 U.S.C. 7501 et seq.).

What does it mean to take public sentiment into account in funding, promoting, adopting, and implementing newly developed technologies, whether this is considered from the point of view of the technologies' developers or its critics? New research and social experimentation will be required to figure this out. We do not actually know the answer to this question at the present time—it is an unsolved problem in democratic practice. We do know that technological choices made today will determine many aspects of our future tomorrow. We do not, in most cases, have adequate and clear mechanisms for formally taking public sentiment into account in such decisions in many major democratic societies, what some have referred to as "public consultation" practices (Einsiedel 2008). Risk communication efforts of the future are likely to be designed to help address this issue, although by now it should be clear that management of risk, narrowly construed, is not the only consideration.

Clearly, market researchers have known for some time that recognition of relevant social values is crucial to the adoption of new alternatives. But for nanotechnology we are talking about alternatives that have benefited by substantial public investment, and that are therefore generally expected to pay back in terms of future benefit. This creates a different dynamic, one in which the relevant ethics are influenced by the level of taxpayer dollars that have been invested. The developers of nanotechnology may be seen as owing a debt to society, in proportion to this investment. But what does society want from technology, then?

One of the useful ways to think about technology, sociologically, is as an *artifact* of culture. Just as stone tools are artifacts of earlier cultures—that is, they are objects modified or created by people to meet their everyday needs—modern technologies reflect our attempts to improve our daily living. Today's automobiles, television networks, video games, computers, clothing, cookware, satellites, subways, ballpoint pens, sporting gear, and cell phones are all artifacts of modern life, things we have made to meet our own needs and serve our own interests, and that will be left here to reflect our history long after we ourselves are gone. It is useful to think about technology this way, because it reminds us that human beings generally create tools and technologies to meet specific human needs; this is an integral part of what makes us human. As such, technology provides a sort of reflection or record of those needs. Technology thus records and embeds both our values and our chosen way of life.

Technology, once adopted, also has a way of constraining our future (Garmire and Pearson 2006). Committed wholesale (in most areas) to automobile-based transportation in local urban environments, it is all the more difficult for the United States to back up and reconsider more investment in alternatives, even relatively modest adjustment such as a shift to electric or hybrid cars, or to alternative fuels. It is easier for us to develop more fuel-efficient alternative vehicles than to tear up the freeways and substitute trains, trams, or subways, even though, in the long term, these might be better alternatives from many perspectives, including with respect to minimizing our contribution to climate change. Of course, this investment in individual automobile transportation is a very clear reflection of human social values—Americans' desire to control their immediate destinies, to be able to go anywhere, anytime, and not depend on someone else's schedule are clearly what has brought us to this point.

Nanotechnology's future will likewise depend on human values. When people are asked to consider future nanotechnology and its risks and benefits, it should not surprise us too much that they may tend to project their more general hopes and fears onto these judgments. People—that is, ordinary citizens, members of broader, nonexpert publics—often seem to understand the role of technology in creating their futures in a remarkably significant way that is directly related to the nature of our culture (that is, to the culture of the developed world, which has benefited in most clear economic terms from contemporary technological advancement). Fortunately for technology's promoters, most people in the developed world tend to begin with the assumption that technology is valuable and will contribute to a higher standard of living, better medical care, and so on. This remains true despite contrary historical experiences where public controversies over certain technologies, sometimes bitter ones, have erupted. We might most accurately understand these as exceptions, rather than standard expectations, for how society will react to emerging technology.

We also all know that environmental and health risks are likely companions of any future class of technology. A recent National Academy of Engineering report (Garmire and Pearson 2006) notes that recognizing the risks as well as the benefits of technology is a basic characteristic of technological literacy. Wise consumers do not dismiss the possibility that technology brings both, and they therefore appreciate opportunities to learn more about both, and to express their opinions on this score. Messages about future benefits will tend to induce consideration of possible future risks; today's consumers are aware of both.

Risk communicators should not be tempted to casually dismiss all public concerns about risks, even where these seem unwarranted, or to blame public ignorance for the emergence of risk-related concerns. Denial of popular concerns is unlikely to lead to their going away. Historical cases in which officials have dismissed public concerns out of hand have often been those that resulted in persistent mistrust and conflict, ranging from the controversies over genetically modified (GM) foods to the association of autism with childhood vaccination.

This is not to suggest that popular thinking about technology is always correct, but that it should always be respected. Trust is hard to gain and easy to lose, and dismissing people's deeply felt concerns, even where these seem scientifically unfounded, is not a good way to gain trust. A better strategy is to acknowledge and consider those concerns. The movement toward upstream engagement and increased public involvement in making decisions about how major technologies and their implications should be managed is based partly on the idea that nonexperts can be invited as rational and thoughtful partners in considering technology's future. It also reflects the recognition that technological development is an expression of societal values. This type of thinking has made nanotechnology an element of a grand experiment in deliberative democracy. So far, the experiment seems largely successful.

Professional risk communicators often represent particular interests, such as those of government agencies, universities, other research organizations, or commercial entities with immediate stakes in the development and deployment of what they see as valuable new technology. When the interests of those organizations that risk communicators are hired to represent seem to clash with concerns of various other publics, trouble is on the horizon. The constructive path forward requires better understanding of nonexperts' points of view. This may require risk communicators to assume and maintain the role of impartial arbiter among competing interpretations of reality. This is a difficult role, and yet they probably should assume it, in the interests both of their clients and of democracy. At present, however, nanotechnology's risks are little discussed in the media and do not seem to have overly pervaded public thinking.

We might best think of nanotechnology as representing *attenuated* rather than *amplified* risks, with respect to popular perception. That is, if anything, public perception may understate, rather than exaggerate, the risks that exist. This is culturally interesting but so far not fully explained. Further, some risks, such as the risk of violation of privacy, that for some reason seem to come to mind among many members of the public when asked about nanotechnology, are simply not high on the list of risks that experts perceive as important (Besley et al. 2008). Do fictional accounts of very tiny robots and miniature cameras proposed for medical tests suggest to ordinary members of the public that nanotechnology will be used to accelerate the development of spyware? We have no direct way to answer this question; it stands as an illustration of how much we have to learn.

However, we do know, based in part on the South Carolina panel study, that ordinary citizens, including members of diverse publics, are capable of reasoned consideration of both the benefits and the risks of nanotechnology. Their expressed opinions tend to be thoughtful and, after an initial period in which risk awareness seemed to rise modestly (Priest et al. in press), remarkably stable (Priest and Greenhalgh n.d.). This contradicts the expectations of some commentators that public opinion about technical subjects is always and necessarily variable and unpredictable, and even fickle. While this could change, as information about both the very real and the imagined risks of nanotechnology diffuses, at present there is no observable tendency for the public to panic or otherwise embrace irrational views.

One possible negative scenario for the future would be the reproduction of the history of agricultural biotechnology and genetically modified food, in which people around the world were shocked to discover that fundamental changes in the way their food was produced had been implemented without their knowledge. Although only one element among many, this idea that things were being done behind their backs which affected what they put into their bodies as food every day was likely an important contributor to public backlash against this technology. Improved opportunities for public engagement should, at a minimum, be able to address this lack of transparency for emerging technology. Other concerns, originally stemming from communities of farmers in various locations around the world, involved unanticipated and uncertain economic impacts that they felt might imperil their livelihoods. Broad public discussion of the distributional justice issues associated with nanotechnology and other emerging technologies might be in order.

Some concerns about biotechnology did not turn out to be fully justified by the best available evidence; others, including long-term economic and environmental impacts, are still open to debate. Either way, a crucial lesson was that transparency and opportunities for public discussion are critical to ensure healthy democratic decision making, moving forward. It is equally important that research on economic uncertainties should accompany expanded research on environmental and health uncertainties, with risk communication efforts, broadly conceived as opportunities for two-way dialogue, mounted alongside both.

An important role for today's risk communication specialists working in nanotechnology and related areas is to make sure information about risks, as well as benefits, is openly available and openly discussed. This is likely to serve the real long-term interests of the technology's promoters just as well as it serves the interests of democracy, a serendipity that should be more widely acknowledged. However, part of this job is likely to involve reminding nanotechnology's champions that this transparency is vital. Scientists and science policy specialists alike should not retreat, in the face of potential controversy, to deficit model interpretations of public understanding, in which public concerns are dismissed to the extent they are attributed to inadequate public understanding of science. Rather, they should seek new opportunities to encourage public discussion of nanotechnology, including its attenuated risks.

PERSPECTIVE PART 1: LESSONS LEARNED FROM THE FIRST U.S. NATIONAL CITIZENS' TECHNOLOGY CONFERENCE[1]

Michael D. Cobb

It is universally accepted that Americans are not very knowledgeable about science (Miller 1998; National Science Board 2010). Yet, many scholars believe that in a democracy ordinary citizens should exert greater influence over the development trajectories of emerging technologies (Kleinman 2000; Sclove 2000; Hamlett 2003; Wilsdon et al. 2005; Powell and Colin 2009). How can citizens effectively participate in policy discussions about complex technologies if they lack a basic understanding of them?

According to much of the science and technology (S&T) literature, the solution to low levels of knowledge is to *engage* citizens with

[1] This research was supported by the Center for Nanotechnology in Society at Arizona State University (NSF grant #0531194). The views expressed in this paper are mine alone and do not necessarily represent those of the National Science Foundation.

issues of science and technology (Kleinman et al. 2009). Engagement involves more than merely disseminating more and more information to the public, which is arguably an ineffective strategy for creating informed opinions (Nisbet and Schuefele 2009). Instead, engagement is organized around creating a dialogue between experts and nonexperts on an equal footing. When citizens are given the opportunity to interact with experts, so goes the argument, they not only exchange equally valuable kinds of information, but citizens also become more efficacious.

In order to promote this kind of dialogue, scholars, interest groups, and even government agencies have organized a plethora of engagement activities, ranging from informal science education at science cafes (Powell and Colin 2009) and science museums (Bell 2008) to intentionally deliberative methods such as citizen juries (Smith and Wales 2000) and consensus conferences (Guston 1999; Einsiedel and Eastlick 2000; Einsiedel et al. 2001; Hamlett and Cobb 2006; Abelson et al. 2007; Powell and Kleinman 2008; Boussaguet and Dehousse 2009; Kleinman et al. 2009; Nelson et al. 2009; Cobb 2011).[2]

In this essay, I briefly describe a particular kind of engagement activity, called the National Citizens' Technology Forum (NCTF). The NCTF was the first U.S. national consensus conference, and it asked citizens to think about technologies that might one day enhance human capabilities. Despite the significant increase in the number and kinds of citizen engagement activities, evaluations of their effectiveness have been slower to materialize. Thus, I summarize some of the key findings of prior analyses of the NCTF, and I present additional results regarding participants' opinions of scientists before and after interacting with them.

My attention to these new data is motivated by two factors. First, scholars warn that citizens might walk away from engagement exercises with a dimmer view of the scientists working in these issue areas (Garvin 2001), so this is an important set of data to analyze. Second, my findings should serve as a cautionary warning that, while the NCTF produced many desirable outcomes, engagement activities attempting to foster deliberation also carry some risks (Mendelberg 2002; Cobb forthcoming).

[2] While creating informed opinions is one goal, many scholars also value engagement forums for their potential to connect citizens' preferences to policy decisions (Petts 2004).

CONSENSUS CONFERENCES AND THE NATIONAL CITIZENS' TECHNOLOGY FORUM (NCTF)

Although a wide variety of public engagement techniques exist to create a dialogue with the public (Smith and Wales 2000; Burgess et al. 2007; Bell 2008; Powell and Colin 2009), consensus conferences are uniquely structured to produce deliberation among participants (Guston 1999; Einsiedel and Eastlick 2000; Einsiedel et al. 2001; Hamlett and Cobb 2006).[3] Opting for depth of discussion instead of the representativeness of participants (e.g., deliberative polling), the number of panelists in each forum typically ranges from 10 to 15. Participants in these events usually meet in a face-to-face setting, and a trained facilitator helps to guide their deliberation. Also, in contrast to other models of engagement that last for hours to perhaps a full weekend, consensus conferences involve a larger number of meetings that can span an entire month.

Other key features of consensus conferences include ensuring access to diverse viewpoints and sources of information, including opportunities for direct contact with scientists and experts. Finally, writing policy recommendations is the crux of the conference. Organizers task participants with the goal of reaching consensus about their recommendations, not with identifying values supported by simple majorities. In principle, the policy reports are disseminated to policy makers. In Denmark, for example, the Danish Parliament's Board of Technology not only distributes the final set of policy recommendations to both the press and the public, but it also gives them to parliament.

The NCTF was the first U.S. national consensus forum, and it was held in March 2008 (Hamlett et al. 2008). Citizens who participated in it were asked to deliberate about *converging* technologies (nanotechnology, biotechnology, information technologies, and cognitive science, or NBIC) to enhance human performance.[4] A few crucial aspects to the NCTF design distinguish it from past consensus conferences. First, as the name implies, the conference included participants from multiple parts of the country. It did so by simultaneously hosting conferences at six different cities representing distinct geographical locations (Atlanta, Georgia; Berkeley, California; Golden, Colorado; Durham, New Hampshire; Madison, Wisconsin;

[3] Ideally, they also connect informed citizen preferences that result from deliberating to decisions being made by policy makers (Macnaghten et al. 2005).

[4] For example, some people speculate that NBIC technologies can one day create medical devices that roam the bloodstream searching for cancer cells.

and Tempe, Arizona). Second, it used the Internet as a mode of communication to manage the challenge of large geographic distances between cities.

Using Internet technology permitted conference-wide deliberative interaction. Procedurally, participants deliberated face-to-face in their respective geographic groups for one weekend at the beginning and at the end of the month, while in between they deliberated electronically across their geographic groups in nine 2-hour sessions. As Delborne and Schneider discuss in this chapter, there is some debate in the literature about the effectiveness of computer-mediated deliberation (Scott 1999; Sunstein 2001; Min 2007), and the NCTF has provided additional data to study its effects (see also Delborne et al. 2011).

Because the conference included six cities that were geographically dispersed, organizers at each of the site locations recruited local citizens to participate using targeted advertisements. Applicants were directed to complete an online survey to measure key demographics of the volunteers and to eliminate industry or interest group partisans. Due to the substantial time commitment, participation was encouraged by providing panelists a $500 stipend that was awarded after completing the consensus conference. While compensation is often used to recruit participants, Delborne and Schneider point out in this chapter that this practice might be problematic (see also Kleinman et al. 2011).

Each site tried to recruit and select citizens with a priority of balancing their socioeconomic characteristics, but strictly speaking, the sample was nonrandom, and with less than 90 participants, it was not representative of the broader U.S. population. Nevertheless, NCTF panelists reflected a reasonable approximation of the American public on several important demographic characteristics (see Hamlett et al. 2008). For example, half the originally invited panelists were women, 65% were white, and the median age and income were, respectively, 39 years old and $50,000 to $75,000.[5] Over 350 people applied to participate, 89 of them were chosen, and 85, or about 15 people per site, attended the first meeting.

The remaining features of the NCTF were routine for consensus conferences. For example, participants received a 61-page background document before attending the first meeting, although most

[5] On the other hand, just 9% of participants self-identified as Republicans, although this figure was nearly identical to the percentage of Republicans who volunteered to participate. Efforts to recruit more Republicans were unsuccessful.

later admitted to not having read it by their first meeting.[6] During the Internet deliberation period, besides deliberating with one another, panelists had question-and-answer sessions with a diverse group of topical experts, including technical specialists, a philosopher, and a specialist in regulatory processes. A professional facilitator made sure that participants had opportunities to speak and ask questions, and intervened if necessary to ensure participants' dialogue remained respectful.[7] Finally, after the last meeting, participants drafted reports that represented the consensus judgments of their local groups.

NCTF OUTCOMES

A few studies about the NCTF have already been published. On the whole, these evaluations indicate that running a complex national consensus forum is achievable (Philbrick and Barandiaran 2009; Wickson et al. forthcoming) and that many positive aspects of the process emerged (Cobb forthcoming). They also indicate that several aspects of the process could be improved, such as how recruitment is conducted and how the Internet is used in these forums (Kleinman et al. 2009; Delborne et al. 2009). Hamlett, Cobb, and Guston (2008), for example, analyzed and compared the six different consensus reports and found an impressive degree of similarities across them. One concern, for example, was centered on equity—participants wanted to know who will get access to enhancements, under what conditions, when they become available (Bal 2011). They also interpreted these reports to be thoughtful and well conceived, and a reasonable basis for lawmakers to use when thinking about future policy decisions.

Yet, as Powell and Colin (2009) observe, it is unlikely that lawmakers will draw upon these recommendations, because formal mechanisms linking the NCTF citizens' recommendations to policy were absent. Similarly, Cobb (2011) finds that participating citizens became more knowledgeable, trusting of others, and internally efficacious, but also that participants reported feeling less externally efficacious. Engagement in the NCTF apparently builds citizens' self-recognized capacities to participate in

---

6 See Anderson et al. (n.d.) for a discussion about the role of information supplied versus independently acquired by participants.

7 Because different facilitators were used at each site to manage the local face-to-face deliberations, variations in the management of deliberation were possible, and this could help explain occasional differences in outcomes across site locations.

decision-making tasks about complex technologies, but it also apparently frustrated participants instead of empowering them without their being able to actually influence policy.

CITIZENS' VIEWS ABOUT SCIENTISTS

Some scholars worry that interactions with scientists will result in undesirable outcomes. Scholars give several reasons to anticipate problems, most of them involving poor communication between laypersons and scientists (Irwin and Wynne 1996; Yearley 2000). As is often cited, nonexperts and experts tend to approach aspects of science and technology from different perspectives (Ho et al. 2010), which can give rise to misunderstandings and eventually alienate citizens from scientific experts (Garvin 2001). If the dialogue between experts and nonexperts alienates citizens rather than enlightens them, this outcome could threaten to undermine the predicted increased efficacy among participants and dampen their enthusiasm about future participation.

In order to evaluate attitudinal outcomes of the NCTF, participants completed lengthy pretest and posttest questionnaires that were taken in addition to the original applicant survey.[8] For the purposes of this analysis, I examine answers to just six questions asking about participants' views of scientists. Specifically, participants were asked whether they agreed or disagreed that most scientists: (1) work for the good of humanity, (2) choose their work to make money, (3) choose their jobs to make things better for the average person, and (4) do not think about whether their products are morally or ethically acceptable to others. In addition to questions about the perceived motivations of scientists, panelists were asked if they agreed or disagreed whether (1) "scientific and technical experts understand the values of ordinary people like me" and (2) "if I were to meet with a scientist in my community, he or she would treat me with respect." Altogether, these kinds of questions are intended to measure, albeit indirectly, whether the kind of citizen–scientist interaction hoped for in engagement models results in more or less favorable opinions.

[8] Data were gathered measuring participants' knowledge and opinions about nanotechnology and human enhancement and their self-reported feelings of efficacy and trust in others. For additional details about the survey methodology and measurement issues, see Hamlett, Cobb, and Guston (2008) and Cobb (2011).

RESULTS

Before deliberating, NCTF panelists held favorable impressions of scientists. A large majority (73%) agreed scientists work for the good of humanity, and most (60%) agreed scientists work to improve things for the average person. Meanwhile, just 7% and 9%, respectively, disagreed with these two propositions, while the remainder reported feeling neutral. Conversely, just 18% and 26% agreed, respectively, that scientists mostly work for money or do not think about the morality/ethics of their activities. Indeed, 44% and 52%, respectively, *disagreed* with these sentiments. Similar to the first four items, majorities (52% and 73%) agreed that scientists understood their values and would treat them with respect. Moreover, not one person disagreed that they would be treated with respect (the remaining 27% answered they were neutral), and just 12% disagreed that scientists understood their values (37% answered they were neutral).

By the end of the NCTF, however, panelists' views about scientists had changed even though they remained positive overall, and the negative direction of the opinion change was statistically significant for three of the six measures. For example, at the end, 60% agreed and 15% disagreed that most scientists work for the good of humanity, a 13% decline in agreement and a doubling of disagreement ($p < .05$). This negative trend was also apparent on the question about scientists working for the average person, where at the end just 45% agreed with this notion and 25% disagreed with it ($p < .05$).

Although there was no real change in the percentage agreeing that scientists mostly work for the money (16% compared to 18%), there was a large *decline* in the percentage *disagreeing* with this statement (29% versus 44%; $p < .10$). On the remaining three items, opinion change was in a more negative direction twice, while opinions did not change on the third, and the two times opinions changed it was not statistically significant. An additional 6%, for example, agreed that scientists fail to consider the ethical dimensions to their work, and now three people (4%) agreed that scientists would fail to treat them with respect.

CONCLUSIONS

One of the hypothesized virtues of citizen engagement with science is that ordinary people become empowered to influence science policy in the future. Although there is some evidence from the NCTF to support this view, such as finding participants' increased knowledge, trust, and perceived internal efficacy, there are also some reasons to be cautious about assuming too much. First, engagement activities can generate undesirable outcomes. My prior analyses of the NCTF, for example, uncovered reduced feelings of external efficacy (Cobb, 2011).

Why do participants become less likely to believe their actions can affect policy outcomes? The logical explanation is the lack of formal linkages between the NCTF and policy decisions, but because there is little demand from policy makers to solicit and incorporate informed public opinion through engagement activities, there is no reason to think future consensus conferences in the United States would lead to different outcomes any time soon (Powell and Colin 2009).

Second, as I find here with citizens' perceptions of scientists, interaction between citizens and scientists might undermine other positive features of engagement. To be sure, participants' views about scientists remained positive at the end, but they clearly became less positive when measured at the end of the NCTF. Of course, I cannot authoritatively demonstrate that poor communication caused citizens to alter their opinions (participants might have rationally adjusted their overly optimistic views about scientists), but these results suggest it is important to pay attention to how citizens and scientists interact to guard against inducing adverse reactions.

Going forward, I think it is imperative to gather more and better evaluative data to determine the merits of engagement. Some forms of engagement might be superior to others on some metrics but inferior to others on different measures. Theory has already been consulted to guide the development of models of engagement, but better empirical measurement of outcomes is necessary to answer these kinds of questions. Future research will surely benefit our understanding of the value of engagement by looking at the long-term effects of it. The presumption of citizen empowerment requires collecting longitudinal data tracking future behaviors and demonstrating that engaged citizens participate more often in policy processes, and that they do so precisely because of their past involvement in activities like the NCTF.

REFERENCES

Abelson, J., P.G. Forest, J. Eyles, A. Casebeer, E. Martin, and G. Mackean. 2007. Examining the role of context in the implementation of a deliberative public participation experiment: results from a Canadian comparative study. *Social Science and Medicine* 64(10): 2115–2128.

Anderson, A.J., J.A. Delborne, D.L. Kleinman, and M. Powell. n.d. Information beyond the forum: motivations, strategies, and impacts of citizen participants seeking information during a consensus conference. Unpublished paper.

Bal, R. 2011. Public perceptions of fairness in NBIC technologies. In *Nanotechnology and the Challenges of Equity, Equality, and Development: Yearbook of Nanotechnology in Society*, Volume 2, Part 3. Eds. S. Cozzens and J. Wetmore. New York: Springer, 231–249.

Bell, L. 2008. Engaging the public in technology policy: a new role for science museums. *Science Communication* 29: 386–398.

Boussaguet, L., and R. Dehousse. 2009. Too big to fly?: A review of the first EU citizens' conferences. *Science and Public Policy* 36: 777–789.

Burgess, J., A. Stirling, J. Clark, G. Davies, M. Eames, K. Staley, and S. Williamson. 2007. Deliberative mapping: a novel analytic-deliberative methodology to support contested science-policy decisions. *Public Understanding of Science* 16: 299–322.

Cobb, M.D. 2011. Creating informed public opinion: citizen deliberation about nanotechnologies for human enhancements. *Journal of Nanoparticle Research*. 13(4): 1533–1548.

Cobb, M.D. Forthcoming. Deliberative fears: citizen deliberation about science in a national consensus conference. In *Thinking Through Technology with Publics*. Ed. K. O'Doherty. Vancouver: University of British Columbia Press.

Delborne, J.A., A.A. Anderson, D.L. Kleinman, M. Colin, and M. Powell. 2011. Virtual deliberation?: Prospects and challenges for integrating the Internet in consensus conferences. *Public Understanding of Science*. 20(3): 367–384.

Einsiedel, E.F., and D.L. Eastlick. 2000. Consensus conferences as deliberative democracy. *Science Communication* 21: 323–343.

Einsiedel, E.F., E. Jelsøe, and T. Breck. 2001. Publics at the technology table: the consensus conference in Denmark, Canada, and Australia. *Public Understanding of Science* 10: 83–98.

Garvin, T. 2001. Analytical paradigms: the epistemological distances between scientists, policy makers, and the public. *Risk Analysis* 21(3): 443–455.

Guston, D. 1999. Evaluating the first U.S. consensus conference: the impact of the citizens' panel on telecommunications and the future of democracy. *Science, Technology and Human Values* 24: 451–482.

Hamlett, P. 2003. Technology theory and deliberative democracy. *Science, Technology, and Human Values* 28(1): 112–140.

Hamlett, P., and M.D. Cobb. 2006. Potential solutions to public deliberation problems: structured deliberations and polarization cascades. *Policy Studies Journal* 34: 629–648.

Hamlett, P., M.D. Cobb, and D. Guston. 2008. "National Citizen's Technology Forum: Nanotechnologies and Human Enhancement." CNS-ASU Report # R08-0002. Retrieved from http://cns.asu.edu/files/report_NCTF-Summary-Report-final-format.pdf.

Ho, S., D.A. Scheufele, and E. Corley. 2010. Value predispositions, mass media, and attitudes toward nanotechnology: the interplay of public and experts. *Science Communication* (published online September 2010. Doi: 10.1177/1075547010380386.).

Irwin, A., and B. Wynne., Eds. 1996. *Misunderstanding Science*. New York: Cambridge University Press.

Kleinman, D.L., Ed. 2000. *Science, Technology, and Democracy*. New York: State University of New York Press.

Kleinman, D.L., J.A. Delborne, and A.A. Anderson. 2011. Engaging citizens: the high cost of citizen participation in high technology. *Public Understanding of Science* 20(2): 221–240.

Macnaghten, P.M., M.B. Kearnes, and B. Wynne. 2005. Nanotechnology, governance, and public deliberation: what role for the Social Sciences? *Science Communication* 27: 268–291.

Mendelberg, T. 2002. The deliberative citizen: theory and evidence. In *Research in Micropolitics: Political Decision Making Deliberation and Participation*. Eds. M.X. Delli Carpini, L. Huddy, and R. Shapiro. Greenwich, CT: JAI Press, 151–193.

Miller, J.D. 1998. The measurement of civic scientific literacy. *Public Understanding of Science* 7: 203–223.

Min, S. 2007. Online vs. face-to-face deliberation: effects on civic engagement. *Journal of Computer-Mediated Communication* 12(4): article 11. http://jcmc.indiana.edu/vol12/issue4/min.html.

National Science Board. 2010. *Science and Engineering Indicators 2008*. National Science Foundation, Washington, DC.

Nelson, J.W., M.K. Scammell, R.G. Altman, T.F. Webster, and D.M.Ozonoff. 2009. A new spin on research translation: the Boston consensus conference on human biomonitoring. *Environmental Health Perspectives* 117: 495–499.

Nisbet, M.C., and D.A. Scheufele. 2009. What's next for science communication?: Promising directions and lingering distractions. *American Journal of Botany* 96: 1–12.

Petts, J. 2004. Barriers to participation and deliberation in risk decisions: evidence from waste management. *Journal of Risk Research* 7(2): 115–133.

Philbrick, M., and J. Barandiaran. 2009. National citizens' technology forum: lessons for the future. *Science and Public Policy* 36(5): 335–347.

Powell, M., and D. Kleinman. 2008. Building citizen capacities for participation in nanotechnology decision-making: the democratic virtues of the consensus conference model. *Public Understanding of Science* 17(3): 329–348.

Powell, M., and M. Colin. 2009. Participatory paradoxes: facilitating citizen engagement in science and technology from the top-down? *Bulletin of Science Technology and Society* 29: 325–342.

Sclove, R. 2000. Town meetings on technology: consensus conferences as democratic participation, In *Science, Technology, and Democracy*. Ed. D.L. Kleinman. New York: State University of New York Press, 33–48.

Scott, C.R. 1999. Communication technology and group communication. In *The Handbook of Group Communication Theory and Research*. Ed. L.R. Frey. Thousand Oaks, CA: Sage, 432–472.

Smith, G., and C. Wales. 2000. Citizens' juries and deliberative democracy. *Policy Studies* 48: 51–65.

Sunstein, C.R. 2001. *Designing Democracy: What Constitutions Do*. New York: Oxford University Press.

Wickson, F., M.D. Cobb, and P. Hamlet. Forthcoming. Review of deliberative processes: National Citizens Technology Forum—USA. In *Democratisation of Science and Technology Development. Deliberate Processes in the Development of Nanotechnologies*. Eds. S. Pål, G. Scholl, and S. Eivind. Singapore: Pan Stanford.

Wilsdon, J., B. Wynne, and J. Stilgoe. 2005. *The Public Value of Science. Or How to Ensure That Science Really Matters*. London: Demos.

Yearley, S. 2000. Making systematic sense of public discontents with expert knowledge: two analytical approaches and a case study. *Public Understanding of Science* 9: 105–122.

PERSPECTIVE PART 2: MOVING FORWARD WITH CITIZEN DELIBERATION: LESSONS AND INSPIRATION FROM THE NATIONAL CITIZENS' TECHNOLOGY FORUM

Jason Delborne and Jen Schneider

In his article on the National Citizens' Technology Forum (NCTF) in this chapter, Cobb notes that the NCTF was essentially a descendant of the "consensus conference," a form of political engagement that originated in Denmark and then traveled elsewhere. Sponsored by the Danish Parliament, the Danish Board of Technology was tasked with involving groups of citizens in making informed policy recommendations related to science and technology: these policy recommendations were and are considered by lawmakers in forming science policy. Cobb and others have noted that the consensus conference and related forms of public engagement have garnered significant academic interest in the United States in recent decades, particularly among those concerned about the lack of public participation in science policy making (Guston 1999; Sclove 2000; Kleinman et al. 2007; Philbrick and Barandiaran 2009; Rowe and Frewer 2009).

A number of scholars have analyzed the significant challenges inherent in attempting to replicate the consensus conference model in the United States. Philbrick and Barandiaran (2009) provide an excellent review of the critical literature in this regard, and in the sections below, we make some further recommendations, garnered from other scholarly literature on the NCTF, for improving the process in its future iterations. However, we wish to emphasize the important role the NCTF plays as a form of potentially meaningful public engagement. Both science and technology studies (STS) and science communication scholarship highlight the importance of this contribution.

Generally speaking, the United States offers few institutionalized contexts for involving citizens with scientists, policy makers, and other relevant stakeholders in making policy for science and technology. Citizens as constituents may certainly involve themselves in science policy, particularly in the case of scientific controversies (see Kleinman et al. 2010), and there are some instances in which citizens may testify at public hearings for particular agencies, such as the Environmental Protection Agency or the Nuclear Regulatory Commission. However, the barriers to meaningful participation in these "downstream" forms of policy making are significant (Cox 2006). Governmental institutions and agencies are slow to adopt new or unorthodox forms of public participation, particularly when they may challenge the status quo.[1]

These institutional barriers to participation reflect historical patterns of science communication and policy making that tend to exclude "lay" knowledge and citizen voices from decision making (Sarewitz 1996; Depoe et al. 2004; Lucena et al. 2010; Schneider and Snieder 2011). In *Managing for the Environment*, O'Leary et al. (1998) provide a succinct history of how science has been communicated to the public in policy contexts. Originally, communicators of the risks and benefits of science and technology saw no difference between sharing information with the public and with experts. However, doing so—often in highly technical terms—resulted in little meaningful engagement with laypersons: communication was one-way and took no account of lay knowledge. Scholars in STS and science communication (e.g., Wynne 1991; Nisbet and Scheufele 2009) refer to this perspective as the *deficit model* (focusing on the

[1] One notable exception is the U.S. Department of Energy, which is pioneering some forms of public engagement in the field of carbon capture and sequestration (see NETL/DOE 2009).

knowledge deficits of the public as the primary obstacle to good science policy), which continues to dominate science policy making and communication.

Science communication and science policy studies have evolved beyond the deficit model. At least in theoretical terms, communication scholars (especially those with an environmental or science and technology focus) now frequently argue that the best and most effective forms of science communication involve meaningful partnerships and engagement with the public (Groffman et al. 2010). Multiple models of public engagement with scientific, technical, and risk decision making exist (e.g., Daniels and Walker 2001; Guston and Sarewitz 2006; Rowe and Frewer 2009). Within this context, the NCTF emerges as a significant experiment in public engagement.

As an experiment, therefore, the NCTF succeeded on multiple fronts. Participating citizens learned a good deal about the converging technologies of nanotechnology, biotechnology, information technology, and cognitive science (NBIC); the final reports generated at the six discussion sites reflected careful thought and some unique insights not typically present in mainstream discussions about the future of NBIC and human enhancement; and project organizers proved their ability to implement an ambitious and complex plan that adapted the consensus conference model to the U.S. context. Beyond these substantive outcomes, however, the dual identities of project organizers as practitioners and researchers created the opportunity for careful analysis of the NCTF and the creation of new knowledge about how to pursue citizen deliberation. As project organizers,[2] we not only had access to the quantitative survey data (discussed by Cobb in "Perspective Part 1: Lessons Learned from the First U.S. National Citizens' Technology Conference," this volume), but also collected qualitative data through participant observation, analysis of transcripts from online sessions, and semistructured interviews with the NCTF participants from the Madison, Wisconsin, site. The remainder of this essay draws upon formal analyses of this diverse mix of data to generate recommendations for refining and improving the model for citizen engagement in the U.S. context.

---

[2] In 2008, Delborne served as the lead organizer and head facilitator of the NCTF in Madison, where he worked as a postdoctoral researcher at the University of Wisconsin–Madison; Schneider collaborated with a colleague at the Colorado School of Mines to organize the NCTF in the Denver region.

Recommendation 1: While monetary incentives may increase the number and diversity of applicants for citizen deliberations, organizers must recognize and account for possible distortions in representativeness.

Unlike an official public opinion poll, citizen deliberations do not claim to access a statistically representative sample of a population; but *representativeness* remains an important mark of credibility for exercises such as the NCTF. Organizers assume, rightly, we think, that decision makers who receive policy recommendations generated by citizen panels will question whether the participants demographically represent their constituents. With this goal in mind, NCTF organizers recruited in multiple media venues in each participating metropolitan area and selected participants to maximize diversity in terms of income, ethnicity, political preference, and highest level of education. In addition, advertisements promised $500 stipends to participants who completed the full exercise.

Kleinman, Delborne, and Anderson (2011) compare the motivations for participation between the NCTF in Madison and a previous Madison consensus conference on nanotechnology, held in 2005. They report that the inclusion of a stipend for the NCTF yielded four times as many applicants as the 2005 effort, which offered only free meals and child care. NCTF survey and interview data confirm that the stipend played a key motivational role for some of the Madison participants, although desire to learn about nanotechnology and participate in research reflect higher average scores on the national survey. Importantly, survey data show a strong correlation with income—a majority of participants earning $50,000 or more annually said that they would have participated without the stipend, while a majority of those earning less than $35,000 said that they would have been unlikely to participate without the stipend.

With some caution, Kleinman, Delborne, and Anderson (2011) note that the 2005 consensus conference (without the stipend) attracted a more "civically engaged" group that "expressed skepticism throughout the consensus conference about the idea that new technologies are automatically good and valuable"; in contrast, the 2008 NCTF attracted participants "more broadly interested in technology" and who saw the NCTF "primarily as a learning opportunity (232)." While stipends alone do not explain this difference—other recruitment practices certainly promoted the two experiences in distinct ways—the contrast encourages attention to the impact of different incentives on the resulting character of the panel.

Stipends may allow less privileged citizens to consider participating in a deliberative exercise—a key benefit in an increasingly stratified and unequal society—but they also may create panels that *represent* the desire to earn money rather than the desire to make a difference. Put another way, finding the demographic mix that matches census figures is insufficient to guarantee representativeness, which is a multifaceted and slippery concept. The recruitment strategies of deliberative exercises necessarily construct a set of values and preferences for the applicant pool that translate into a panel that cannot represent the idealized "blank slate" of citizens for rational and unbiased deliberation (for further discussion on the problematic goal of recruiting "ordinary citizens," see Powell et al. 2011).

Recommendation 2: Internet-based citizen deliberations hold some promise for reducing the cost of organizing geographically diverse participants, but the trade-offs and logistical details deserve careful consideration.

Delborne et al. (2011) conducted a thorough analysis of the online component of the NCTF, seeking to explain why participants became relatively disillusioned with the "keyboard-to-keyboard" (K2K) parts of the deliberation. First, the authors found that the specific online interface chosen for the NCTF—a private and moderated chat space with some extra communication features—failed to promote a high degree of coherence for the virtual deliberation. In particular, participants found it difficult to follow multiple discussion threads that appeared in chronological, rather than thematic, order.[3] Second, while any citizen deliberation requires strong facilitation and thoughtful structure, the K2K component of the NCTF promoted little autonomy among participants. As an inductive study of the "norms of deliberation" demonstrated, "Being able to criticize the rules, change them, and change the agenda created participant 'buy-in,' while facilitator control of these matters made the participants passive" (Mansbridge et al. 2009; also see Noveck 2004). During the initial K2K session of the NCTF, participants were unable to negotiate changes in discussion format, anticipating difficulty in staying engaged during online sessions when they were not "chat active."[4] Interview data revealed disturbing levels of disengagement by some Madison participants, especially when they were not "chat active." Consequently, many of them became frustrated with the process, and at least two-thirds exhibited multitasking behaviors during K2K sessions, which went unnoticed by moderators who had

[3] All NCTF K2K transcripts are available at http://www4.ncsu.edu/~pwhmds/online_session_notes.html.

[4] In order to avoid overwhelming the chat space, organizers assigned participants to different groups, each of which would hold "chat active" status during only some portions of the K2K sessions.

few tools to monitor the attention of NCTF participants deliberating from their home computers.

Delborne et al. (2011, 379–381) offer several suggestions to improve future virtual deliberation exercises. First, online deliberation could take advantage of a mixture of *synchronous* and *asynchronous* communication. A standard chat experience permits conversation-like exchanges in real time (synchronous), but an online interface that resembles a bulletin board or wiki would allow participants to offer more considered and organized contributions and ideas (an asynchronous exchange—participants need not be logged in at the same time to share thoughts and information). Second, specific improvements to the online interface would enhance the quality of deliberation: different types of communication could appear in different windows (e.g., logistical reminders, moderator instructions, comments on a particular question or theme); the program could allow moderators to "stack" discussions, both to organize discussion threads by topic and to encourage contributions from quieter participants; and increasing computing capabilities offer the possibility to integrate video and audio into online deliberations, especially during conversations between citizen participants and experts. Third, the facilitation plan could better segregate deliberative tasks that work well in a virtual environment versus those that function better in a common physical space. Question and answer sessions with experts, for example, can work very well online, especially because they offer participants an easy way to consult such information in later stages of deliberation; but decision making, for example, benefits from a more personal connection. Finally, we would also note that trends toward increased online communication, especially among younger generations, make online deliberation increasingly attractive and functional; the NCTF provides us with significant data to improve future iterations.

Recommendation 3: Especially as parts of deliberative exercises move online, organizers need to pay greater attention to participants' access to and search for information beyond that provided formally.

Anderson et al. (n.d.) explore the patterns of "information seeking" among NCTF participants and find an impressive diversity of motivations for and methods of informal research. Especially because the NCTF occurred over a period of weeks, participants conducted Google searches, read and shared media articles, and followed up on sources suggested by experts during online forums. In addition, some Madison participants shared the results of research on an e-mail listserv, requested by the citizens and set up by the lead facilitator. These behaviors show how citizen deliberation draws upon sources of information well beyond vetted background materials and formal conversations with recruited experts.

Even though it is tempting to assume that "more information is good information," Anderson et al. (n.d.) note some concerns with the patterns of information seeking by NCTF participants. First, differences in the ability to conduct research (whether intellectual, technological, or motivational) likely impact the power of different panelists to argue their perspectives. For fact-based arguments, this may be totally appropriate, but many discussions during deliberative exercises reveal conflicts of values. Second, there is little "quality control" of information sought by participants on their own time. Google proved a popular resource among NCTF participants, but Internet searches yield a mix of credible and incredible sources, especially on highly controversial topics such as nanotechnology.

Anderson et al. (n.d.) suggest a number of strategies for incorporating participant information seeking into the design of future deliberative exercises. Most importantly, organizers can explicitly recognize the likelihood and benefit of participants conducting their own research. In doing so, they can facilitate conversations among the panelists about strategies to access diverse sources of information and assess their relative quality. Institutionalized methods for sharing outside information (e.g., a wiki created for citizen panelists) would not only encourage thoughtful information seeking, but also would promote the collective processing of information that complements (and perhaps contradicts) the provided background materials and the perspectives provided by official experts. Such coordination of intragroup communication outside the bounds of facilitated deliberation may also promote a sense of community and investment in the deliberative task—outcomes that increase the motivation for participants to commit their energy, time, and patience to the challenges of citizen deliberation.

The inspiration and rationale for the NCTF remain central to the vision for greater democratic governance of emerging technologies such as nanotechnology. Critiques of the deficit model of the public understanding of science reveal the potential for lay citizens and experts to communicate and learn together; deliberative exercises create opportunities to transform a mix of information, perspectives, and values into thoughtful policy recommendations. The NCTF reveals these benefits but also inspires reflection and revision. As a result of research conducted on the NCTF as a case study of citizen deliberation on nanotechnology, future organizers should pay particular attention to citizen motivations, the potentials and pitfalls of online deliberative components, and the information-seeking practices of participants.

ACKNOWLEDGMENTS

Research for this project was supported by the National Science Foundation under cooperative agreement #0531194. Any opinions, findings, and conclusions are those of the authors and do not necessarily reflect the views of the National Science Foundation or the coordinating institution, the Center for Nanotechnology in Society at Arizona State University.

REFERENCES

Anderson, A., J. Delborne, D. L. Kleinman, and M. Powell. Information beyond the forum: motivations, strategies, and impacts of citizen participants seeking information during a consensus conference. *Manuscript under review*.

Cox, R. 2006. *Environmental Communication and the Public Sphere*. Thousand Oaks, CA: Sage.

Daniels, S.E., and G.B. Walker. 2001. *Working through Environmental Conflict: The Collaborative Learning Approach*. New York: Praeger.

Delborne, J.A., A.A. Anderson, D. L. Kleinman, M. Colin, and M. Powell. 2009. Virtual deliberation? Prospects and challenges for integrating the Internet in consensus conferences. *Public Understanding of Science OnlineFirst* (October 9). doi: 10.1177/0963662509347138.

Depoe, S.P., J.W. Delicath, and M.A. Elsenbeer. 2004. *Communication and Public Participation in Environmental Decision Making*. New York: State University of New York Press.

Groffman, P.M., C. Stylinski, M.C. Nisbet, C.M. Duarte, R. Jordan, A. Burgin, M.A. Previtali, and J. Coloso. 2010. Restarting the conversation: challenges at the interface between ecology and society. *Frontiers in Ecology and the Environment* 8(6): 284–291. doi: 10.1890/090160.

Guston, D.H. 1999. Evaluating the first U.S. Consensus Conference: the impact of the citizens' panel on telecommunications and the future of democracy. *Science Technology Human Values* 24(4): 451–482. doi: 10.1177/016224399902400402.

Guston, D.H., and D. Sarewitz. 2006. *Shaping Science and Technology Policy: The Next Generation of Research*. Madison, WI: University of Wisconsin Press.

Kleinman, D.L., J.A. Delborne, and A.A. Anderson. 2009. Engaging citizens: the high cost of citizen participation in high technology. *Public Understanding of Science* 20(2): 221–240.

Kleinman, D.L., J.A. Delborne, K.A. Cloud-Hansen, and J. Handelsman, Eds. 2010. *Controversies in Science and Technology: From Evolution to Energy*. Vol. 3. Science and Technology in Society. New Rochelle, NY: Mary Ann Liebert. www.liebertpub.com/products/product.aspx?pid=374.

Kleinman, D.L., M. Powell, J. Grice, J. Adrian, and C. Lobes. 2007. A toolkit for democratizing science and technology policy: the practical mechanics of organizing a consensus conference. *Bulletin of Science, Technology and Society* 27(2): 154–169. doi: 10.1177/0270467606298331.

Lucena, J., J. Schneider, and J.A. Leydens. 2010. *Engineering and Sustainable Community Development* 5(1).

Mansbridge, J., J. Hartz-Karp, M. Amegual, and J. Gastil. 2009. Norms of deliberation: an inductive study. *Journal of Public Deliberation* 2(1) (July 27): Article 7.

NETL/DOE. 2009. *Best Practices for Public Outreach and Education for Carbon Storage Projects*. National Energy Technology Laboratory, Department of Energy, December. http//www.netl.do.gov/technologies/carbon_seq/reshelf/BPM_PublicOutreach.pdf.

Nisbet, M.C., and D.A. Scheufele. 2009. What's next for science communication? Promising directions and lingering distractions. *American Journal of Botany* 96(10). doi: 10.3732/ajb.0900041.

Noveck, B.S. 2004. Unchat: democratic solution for a wired world. In *Democracy Online: The Prospects for Political Renewal through the Internet*. Ed. Peter M. Shane. New York: Routledge, 21–34.

O'Leary, R., R.F. Durant, D.J. Fiorino, and P.S. Weiland. 1998. *Managing for the Environment: Understanding the Legal, Organizational, and Policy Challenges*. San Francisco: Jossey-Bass.

Philbrick, M., and J. Barandiaran. 2009. The National Citizens' Technology Forum: lessons for the future. *Science and Public Policy* 36(5): 335–347.

Powell, M.M. Colin, D.L. Kleinman, J. Delborne, and A. Anderson. (2011). "Imagining Ordinary Citizens? Conceptualized and Actual Participants for Deliberations on Emerging Technologies," *Science as Culture* 20(1): 37–70. First published on: 19 August 2010 (iFirst) DOI: 10.1080/9505430903567741.

Rowe, G., and L.J. Frewer. 2009. A typology of public engagement mechanisms. *Science Technology Human Values* 30(2)(July 27): 251–290. doi: 10.1177/0162243904271724.

Sarewitz, D. 1996. *Frontiers of Illusion: Science, Technology and the Politics of Progress*. Philadelphia: Temple University Press.

Schneider, J., and R. Snieder. 2011. Putting partnership first: a dialogue model for science and risk communication. *GSA Today* 21(1) (January): 36–37.

Sclove, R.E. 2000. Town meetings on technology: consensus conferences as democratic participation. In *Science, Technology, and Democracy*. Ed. D.L. Kleinman. Albany: State University of New York Press,33–48.

Wynne, B. 1991. Knowledges in context. *Science, Technology, and Human Values* 16(1) (January 1): 111–121.

References

21st Century Nanotechnology Research and Development Act (15 U.S.C. 7501 et seq.).

Besley, J.C., V.L. Kramer, and S.H. Priest. 2008. Expert opinion on nanotechnology: risks, benefits, and regulation. *Journal of Nanoparticle Research* 10(4): 549–558.

Einsiedel, E.F. 2008. Public engagement and dialogue: a research review. In *Handbook of Public Communication on Science and Technology.* Eds. M. Bucchi and B. Smart. London: Routledge, 173–184.

Garmire, E., and G. Pearson. 2006. *Tech Tally: Approaches to Assessing Technological Literacy.* Committee on Assessing Technological Literacy, National Academy of Engineering, National Research Council. Washington, DC: National Academy Press.

Habermas, J. 1985. *The Theory of Communicative Action: Reason and the Rationalization of Society.* Trans. T. McCarthy. Boston, MA: Beacon Press.

Nordmann, A. 2004. *Converging Technologies—Shaping the Future of European Societies.* Report, European Commission, Brussels.

Pearson, G., and A.T. Young. 2002. *Technically Speaking: Why All Americans Need to Know More about Technology.* Washington, DC: National Academies Press.

Priest, S., and T. Greenhalgh. n.d. Nanotechnology as an experiment in democracy: how do citizens form opinions about technology and policy? *Journal of Nanotechnology Research.* Paper (to be) presented at the Society for Nanotechnology and Emerging Technologies Annual Meeting, Arizona State University, October 2011.

Priest, S., T. Greenhalgh, and V. Kramer. 2010. Risk perceptions starting to shift? U.S. citizens are forming opinions about nanotechnology. *Journal of Nanoparticle Research* 12(1): 11–20.

Priest, S., T. Lane, T. Greenhalgh, L.J. Hand, and V. Kramer. In press. Envisioning emerging nanotechnologies: a three-year panel study of South Carolina citizens. Unpublished paper.

Roco, M.C. 2003. Broader societal issues of nanotechnology. *Journal of Nanoparticle Research* 5(3): 181–189.

Rogers-Hayden, T., and N. Pidgeon. 2007. Moving engagement "upstream"? Nanotechnologies and the Royal Society and Royal Academy of Engineering's inquiry. *Public Understanding of Science* 16(3): 345.

Vidal, J. 2010. Academic resigns from U.K. food watchdog over 'GM propaganda.' Guardian.co.uk. www.guardian.co.uk/environment/2010/jun/02/gm-food-public-dialogue.

Wright, S. 1994. *Molecular Politics: Developing American and British Regulatory Policy for Genetic Engineering, 1972–1982.* Chicago: University of Chicago Press.

6

Audiences, Stakeholders, Cultures, and Nanotechnology Risk

Research continues to raise concerns that nanoscale materials could have negative effects on both human health and the environment. Those individuals who work with nanoparticles in laboratories and factories may be at special risk, and appropriate safeguards remain under development. At present, there appears to remain more uncertainty than certainty about how important these risks might be and how best to protect those exposed. However, nanoparticles are small enough to pass through cell membranes, and some have speculated that these particles may therefore be able to cross the blood–brain barrier and enter the brain, with unknown consequences. In 2009, news reports from China suggested that nanoparticle exposure was responsible for two worker fatalities in a poorly ventilated Chinese plant (Song et al. 2009), although toxicology experts have questioned whether the nanoparticles could really be determined to be responsible, given that the workers in question were probably exposed to many different chemicals. The case illustrates our general lack of knowledge, even if it turns out nanoparticles were not the culprit, and reminds us that just as nanotechnology is a global development, the regulatory challenges are global in nature as well.

Studies of the potential effects of carbon nanotubes on lung tissue are perhaps particularly unsettling. In May 2008, research published in *Nature Nanotechnology* by Poland et al. (2008) showed that some types of carbon nanotubes may behave somewhat like asbestos in the body, potentially leading to the lung cancer known as mesothelioma. Over 2 years later, in September 2010, the U.S. Environmental Protection Agency (EPA) issued new rules under the Toxic Substances Control Act to require notification of intent to use carbon nanotubes in new ways, citing the existence of concern about potential health and environmental hazards. Nevertheless, and while this is an important step, policy for the regulation of nanotechnology's health risks is emerging only slowly and remains in its earliest stages. This is not necessarily attributable to reluctance on the part of government officials to regulate these new products, but is also a matter of the level of resources required, the need to ground regulations in laws that were not designed to cover nanotechnology and its products (a class defined primarily by scale, not specific chemical properties), and the ongoing lack of a solid foundation of research evidence to guide regulatory decision making.

Other forms of nanotechnology pose primarily environmental concerns. According to an April 29, 2010, presentation by a U.S. Environmental Protection Agency (EPA) official, as reported on the Web site of the Woodrow Wilson Center Project on Emerging Nanotechnologies, an estimated 600 unregistered nanosilver products are marketed in the United States, with manufacturing of these products spread around the world (Jordan 2010). Nanosilver is believed to have antimicrobial effects, and these products are often marketed based on this property. The consumer items in question range from cleaning sprays to food containers and from toys to washing machines, all of which represent applications where the ability to kill microbes is seen as desirable. Exactly what engineered nanoparticles such as nanosilver might do to the environment (e.g., in aquatic environments, to fish, to helpful bacteria, or to drinking water) is difficult to project. Clearly not enough is known about these potentialities either, and regulatory action on the environmental front has generally been slow as well. The International Center for Technology Assessment and other organizations petitioned the EPA in 2008 to take action to regulate these nanosilver products. One avenue that is emerging is the application rules designed to cover pesticides, which will be applied to this particular class of products.

In 2009, the U.S. National Institute for Occupational Safety and Health (NIOSH), the U.S. federal agency responsible for doing the research needed to assure worker health and safety, issued a report on "Approaches to Safe Nanotechnology." The document provides interim recommendations and calls for future information exchange to ensure that worker safety and health are protected. But the 104-page document is described on the NIOSH site as "only a starting point" (NIOSH 2009). With methods for evaluating the safety of particular forms of nanomaterials still under development, even the experts are often uncertain where to begin. As we very clearly learned from the case of genetically modified (GM) foods and other agricultural biotechnology, preexisting regulatory structures, including the "division of labor" between various regulatory agencies, generally need to be adjusted when a new set of unanticipated risks comes along (Kuzma and Priest 2010). This can be a complex journey along a path that is not well charted. Novel risks are often poorly understood, presenting a further complication. And bureaucracies are notorious for being slow to change. Even though this may not always be a fair criticism of particular offices or agencies, modern societies are governed by complicated organizations that can be difficult to redirect to meet new problems.

Much of the time, risk messages are aimed at consumers, but they can also be aimed at political leaders (e.g., members of Congress and of parliaments), as well as bureaucratic officials. Consumer activist groups might want to raise public awareness of certain risks, but they can also seek to stimulate regulatory action. Other stakeholders may want to deflect attention from these risks or downplay them to avoid what they consider excessive or unnecessary regulation. Under some circumstances, industry leaders may oppose regulation they consider an unnecessary impediment to doing business. Regulating a risk can legitimize popular concerns and might be seen as fueling risk amplification. On the other hand, commercial and industrial interests can also be concerned when regulation is unclear or insufficient, because this creates uncertain market conditions and unknown business liabilities. Self-regulation may seem an attractive option, but it is unlikely to put consumers at ease. Just how much regulation is too much, and how much government oversight is excessive, are often matters of bitter debate—often, the answers depend as much on regulatory philosophy and ideology as they do on scientific evidence about risk.

In order for governmental agencies to create new rules for new technologies, not only do existing regulations need to be reinterpreted and new regulations written, approved, and then implemented, but credible research on risks generally has to materialize, and scientists must reach at least some level of consensus on what is safe and what is not. This is particularly true in the United States, due in part to its "free market" assumptions (that is, the idea that business thrives in an environment of limited regulation, and that the market will correct errors). However, even in Europe, where there is greater influence of the "precautionary principle" calling for proactive regulation without clear evidence of harm until safety can be assured, regulation does not happen instantaneously and may be unlikely while the relevant scientific evidence is still in doubt. Further, the bureaucratic infrastructure needed for environmental and health regulation of industry may be especially lacking in the areas of the world that have historically lagged behind in terms of economic development; in some respects, such regulations are a luxury of the developed world. Yet nanotechnology is a global development.

These necessities plus the reality of resource constraints faced by researchers, regulators, and policy makers to investigate new issues can certainly slow government response to the risks of new technologies. The complexities of today's globalized markets and international trade agreements, along with the degree of variation from one country to another in development and philosophy of regulatory systems, all introduce further complexity. Nanotechnology is a good example, as international developments in defining acceptable risks, developing standards, and implementing regulatory strategies continue. Without adequate and stable regulation, the benefits of new technology can be lost or delayed because both investors and consumers may become overly cautious. Regulation does not happen automatically; it needs to be initiated, shaped, and managed by human actors as new technologies emerge. Nanotechnology not only represents a new frontier in material science and engineering, but also a new frontier in understanding toxicity. The resources of the risk society are strained by the rapidity of technological change in this as in many other areas.

There are very good reasons why consumers need to understand the risks of the products they purchase, but they are not the only audience or "public" for messages about riskiness. In addition to reaching consumers, who are at this point often "on their own" in making marketplace choices about products claiming benefits from nanotechnology, risk communication often involves getting regulators and policy makers to evaluate risks and take action, as well as getting both industry and advocacy groups to cooperate in this effort. Policy think tanks, consumer-interest organizations, and the donor organizations that fund their activities, alongside government officials and industry leaders, are important actors on the risk policy stage, without which regulatory action by state agencies would likely be even slower to materialize.

The Role of Advocacy in Democracy

In short, it is not just consumer-citizens considering nanotechnology-enhanced products who are the targets of risk communication messages. Under some circumstances, it is policy makers, politicians, rank-and-file regulators, agency officials, and industry leaders who are also targets of risk communication messages, as well as sources of those messages, in some cases. Where applicable, important risk messages may need to reach workers, including researchers and research workers, and those who employ them, as well as labor organizations acting on their behalf. Other identifiable stakeholders for emerging nanotechnology range from the investors who fund start-up companies working with this new technology to medical professionals who ultimately hope to provide new treatments, and their patients; research organizations (including universities) hoping to use nanotechnology to advance their profiles; farm managers who may be interested in using nanobiotechnology fertilizers or pesticides in agriculture; and many others. Environmental resource managers should also be interested in nanotechnology and its risks, both as a source of remediation alternatives and because of potential environmental harm. Each of these groups is also an audience for risk communication messages.

Finally, the taxpayers who are paying the bill for the development of the basic science and engineering underlying the development of nanotechnology are also stakeholders. Approximately $1.8 billion has been proposed to be spent in U.S. federal dollars alone in the president's 2011 nanotechnology budget supplement, about the same as 2010, with many states contributing as well (e.g., by constructing nanotechnology laboratories or by funding nanotechnology research at state universities). In an oft-quoted prediction, the U.S. National Science Foundation has projected that by 2015 nanotechnology would have a $1 trillion impact on the world's economy. Even if this projection is unrealistic, it serves to illustrate the vision of nanotechnology's potential held by its promoters at the highest level. The stakes are very high. A set of technologies representing this kind of investment and with this kind of potential economic impact could be derailed if public opinion should reject it out of hand; on the other hand, if health and environmental risks are not appropriately managed, then derailment on other grounds is one possible outcome.

So far neither scenario seems to be on the horizon, even though risk awareness may be rising. Nanotechnology is being developed in a context influenced by the controversies that emerged for biotechnology in past decades, and thus much attention has been devoted to appropriate communication and management of its possible risks. Yet investment in research on the risks and the development of appropriate risk communication strategies has not entirely kept pace. The U.S. National Nanotechnology Initiative has estimated that $68 million was spent on risk-related research in 2006, although the Project on Emerging Nanotechnologies claims it could only find $13 million being spent in that year on research they deemed "highly relevant" to risk (Maynard 2008). Investment in risk communication research and practice is very difficult to estimate; however, some governmental agencies have funded research in these areas. The U.S. National Science Foundation funded research on nanotechnology has included projects on health, safety, and environment ($74.5 million in 2009 and $91.6 million in 2010), as well as research on education and societal dimensions ($36.8 million in 2009 and $39.9 million in 2010) that includes projects aimed at improving both risk communication and public understanding (National Science and Technology Council Committee on Technology 2010).

Little or none of this investment would be directed specifically toward creating messages aimed at policy makers; in the United States, federal dollars cannot be used directly for lobbying for change, including advocating for responsible regulation. As general attention to nanotechnology has increased along with research investment, awareness of risks as well as benefits among key policy makers will have increased as well, and general public education efforts (while unlikely to focus specifically on risk–benefit trade-offs) reach many audiences. Nevertheless, advocacy efforts are not funded by research dollars, even research dollars spent on investigating societal implications. In order to advocate that regulators, policy makers, and other leaders pay more attention to risks and their regulation, private funding is essential. Enlightened nanotechnology businesses should also consider participating in such efforts, as well as efforts at self-governance. Arguably, it can serve their interests to help create legislation they can live with to guide their management of risks.

What kind of process for developing effective regulation of risks best serves the interests of society as a whole is an open question. In pluralistic democracies with decentralized power and in which a variety of ethnic, cultural, religious, and political groups coexist, such as the United States or the European Union, the interests of various groups are usually seen as competing against one another, and a common assumption is that as no one group can readily dominate this competition, everyone's interests will be served. The system, in other words, may be assumed to guarantee that the interests of the full range of participating groups will be represented. However, this competition has historically produced unbalanced and sometimes unjust outcomes with respect to managing risk. The risk exposure of people in the less-developed world may be sidelined in comparison to the exposure of people in the developed world who have both more risk awareness and more political and economic power. Even within developed societies, the environmental justice movement has uncovered an endless stream of cases in which poorer neighborhoods or those with more minorities have been given an unfair burden of environmental and health risk, as when pollution-producing industries are located in the least affluent neighborhoods (Cole and Foster 2001).

Public advocacy with respect to risk interventions is also unevenly distributed. For example, at any given moment, one set of advocates—let's say women concerned with breast cancer risk—may do better at gaining public sympathy and research dollars than another set of advocates—let's say men concerned with prostate cancer risks. Such groups may have differential access to resources, including media coverage, and therefore more effective power than others. Of course, in this example, both of these diseases deserve increased attention and investigation. But what would be the reasonable, right, and just distribution of risk research dollars between the two? Could this perhaps be resolved by looking at the number of cases or the number of deaths? In the United States in the year 2010, there were about 209,000 new cases of breast cancer and approximately 40,000 deaths (American Cancer Society 2010). There were about 218,000 cases of prostate cancer and around 32,000 deaths. Yet these figures do little to suggest how much public money should be spent on research or prevention for each of the two diseases. Other cancers, such as pancreatic cancer, which had a roughly comparable number of 2010 deaths at 37,000, seem to receive relatively little public attention. A quick Internet search turns up 63.7 million results for breast cancer, only 13.6 million for prostate cancer, and just 6.5 million (about a tenth as much as for breast cancer) for pancreatic cancer.

Factors such as the perceived feasibility of finding new cures, the availability of promising lines of research, the interests of the health research community, including those of commercial entities, and many additional factors other than incidence or mortality are involved in the messy process of prioritizing risks. The amount of time and money spent on risk-related advocacy is also undoubtedly a major factor. Many "orphan" diseases strike such a small number of people that pharmaceutical companies hoping to develop marketable cures are not able to invest research dollars into these particular illnesses. Only through the actions of advocates do these less common risks receive attention and research funding. The playing field, in other words, is not even.

Government attention and action on matters of risk does not happen automatically but is subject to many social and political influences, with power unequally distributed among the various stakeholders. Like other aspects of technology and like other forms of social change, risk regulation evolves in response to societal pressures, which are in turn responsive to both the distribution of power and to underlying societal values. Although industry may sometimes operate better in a regulated environment, in many arenas individual consumers can also resist change in the direction of reducing risk. We have all met people who drive automobiles without seatbelts or while talking on cell phones. Neither governmental and industrial leaders nor researchers are exempt from the factors that sometimes cause people to ignore risks. Public awareness of risk is important but may not be enough to ensure public action, let alone public safety.

It is hoped that in the marketplace of ideas, the existence of a variety of voices will help give balanced consideration to a variety of risks. Yet, in the struggle among various interest groups and other stakeholders to get their risk messages across and increase public attention to their concerns, a lot of opportunity is created for risks to become either overstated (amplified) under some circumstances, or understated (attenuated) under others, in the process. Just which risks are being overstated and which understated at a given moment is rarely clear at the time, however, and may remain largely a matter of opinions, driven in part by values and priorities and not only by risk science. Particular cultures, and particular individuals, are sometimes described as being more "risk averse" than others, meaning they seem to have a lower tolerance for any risks, but the real picture is more complex. Nanotechnology, as a socially attenuated risk rather than a socially amplified one (to date), helps illustrate just as well as biotechnology did that it is not just the scientific elements of a risk that are difficult to understand and predict—it is also, on many occasions, the social ones.

Social, Cultural, and Psychological Influences

One of the big lessons of nanotechnology for risk analysts and risk communicators alike is the strong reminder and confirmation of something we already knew: responses to risks are social, cultural, and psychological in nature. And while individual responses to risks can certainly reflect individual personality characteristics and personal experiences, broader factors are also at work that affect the synergy of people thinking and acting collectively, reflecting the existence of shared perceptions, values, and beliefs (culture), as well as taking place within a network of social groups and institutions (social structure). The social amplification (or attenuation) of risk is a result of processes of interaction that go beyond the particular actions, opinions, and attitudes of individuals, in other words. Further, these social and cultural influences are never likely to be limited to measurable probabilities and other scientific information. Rather, broadly shared values and specific institutional structures and priorities—all of which can be remarkably resistant to change—powerfully influence societal responses to technological and other risks, as well as the desire for certain benefits.

One of the earliest serious challenges to the *technocentric* model of risk analysis underlying conventional technology assessment was the work of social psychologist Paul Slovic, who demonstrated that a variety of psychological factors consistently influence judgments about risk (Slovic et al. 1981; for an overview of his work as it subsequently developed over several decades, see also Slovic 2000). These factors include whether a risk involves the potential for large-scale catastrophic events, whether the risk is familiar or unfamiliar, whether scientists are believed to understand the risk, and so on. Because one goal of this approach to the psychology of risk has been meaningful and accurate measurement of these factors, it has since been termed the "psychometric paradigm" (Siegrist 2010). Another well-known work in this area includes that of Tversky and Kahneman (1986), who were among those researchers who showed that people take various shortcuts or "heuristics" in assessing risk.

Social constructionism—the sociological paradigm that understands human perception of reality as deeply rooted in social learning and other social and cultural factors (Berger and Luckmann 1991)—is generally consistent with the psychometric paradigm, although Slovic has been criticized for not always making a clear distinction between risk as perception and risk as real-world probability (Bradbury 1989). To a strict social constructionist, perception literally *is* reality, whereas many risk theorists, even if not specifically adopting a technocentric view, still tend to see perception as a *distortion of* reality. Some theorists point out that this thinking tends to privilege expert views, which are assumed to be correct and form the standard against which the views of ordinary people are often judged. In actual situations where either individual or public policy decisions about risk are being made, however, the actual risk is unknown, and even among experts there is much room for error. Further, risk perception always involves values, not just scientific facts. Thus, using expert or scientific views as the standard against which the perceptions of others should be judged is, from a social constructionist point of view, a questionable practice.

With respect to risk perception for nanotechnology, it should not be too surprising that nonspecialists appear to see nanotechnology as relatively less risky, in some respects, than some forms of biotechnology. People may not see nanotechnology as posing much threat of catastrophe, they may feel that they have control over whether they personally are exposed to nanotechnology risks, and they may believe that even though nanotechnology is unfamiliar to them, scientists and engineers probably understand these risks sufficiently well to manage them appropriately. It is possible to speculate that if these beliefs are challenged—if the uncertainty surrounding nanotechnology risks even among experts becomes more widely recognized, and if consumers become more fully aware of the increasingly common use of nanotechnology in everyday products ranging from socks to skin cream to skis—public complacency may evaporate. But so far this has not happened, and it may well be that the particular social factors that tended to make people cautious about biotechnology are not operating, for whatever reason, for nanotechnology.

One potential explanation is that our existing paradigm for understanding how people understand risks is incomplete. The psychometric paradigm was developed in part to explain public perceptions of nuclear power risk, where the predominant vision among many opponents involved concern about potentially catastrophic outcomes, concern that is understandable whether or not we favor the expansion of nuclear power. Agricultural biotechnology challenged basic cultural values such as the idea that our food supply should be reliably pure and natural and that the independent family farm should be preserved, uniting those who would like to see small-scale farming, including organic approaches, thrive with others who simply feel it is wrong to go against Mother Nature. Stem cell research has been opposed by those with fixed positions about the moral status of the embryo (Pardo and Calvo 2008), while the prospect of human cloning may have threatened our very sense of self-identity. Nanotechnology, while it carries very tangible risks in some forms, does not seem to threaten catastrophe and does not necessarily seem to challenge most ordinary people's sense of ethics. Risks may become socially amplified for a variety of reasons, but amplification is not inevitable unless our sense of security or other deeply held values are at stake.

This basic thesis—the idea that deeply held cultural values and beliefs are the driver of risk amplification processes—is lent support by looking at the ways in which nanotechnology *is* seen as risky by nonexperts, in particular its perceived risk to privacy (Cobb and Macoubrie 2004; Priest and Greenhalgh n.d.). This particular perception is not related to risk in a scientific sense but makes perfect sense if we remember that privacy is a deeply cherished right for most Americans and many other peoples. It is implied by various amendments to the U.S. Constitution and respected in the laws of many other countries, particularly in provisions restricting the rights of the government with respect to information about individual citizens and those citizens' freedom to engage in a variety of protected (and private) activities without interference. Even though experts and other observers may find scenarios envisioning that governments will use nanotechnology to spy on people rather unlikely, this prospect clearly disturbs people. The idea that the benefits (as well as the risks) of nanotechnology may be distributed unequally also challenges many people's basic values, a concern regularly reflected in interview-based studies (e.g., Priest et al. in press).

Some scholars, notably Douglas and Wildavsky (1983), attempted to further develop this notion that our response to perceived risks and dangers is rooted in cultural factors; they conclude that risk is best conceptualized as a "collective construct." Culture is a collective phenomenon, one that psychological studies focusing on individual perceptions may have a difficult time reflecting. After all, it is clear that the attractions of some technologies are also rooted in culture, with the freedom provided by the individual automobile and the family autonomy implied by individual home ownership particularly cherished by Americans, even when great financial sacrifice and environmental costs are entailed. So why should this idea not apply equally to novel risks—and our choices about which risks are tolerable and which are not—as well?

Recognizing the cultural nature of reactions to risk, some risk communication and public engagement specialists, such as U.K. scholar Brian Wynne, have stressed the collective wisdom that lies outside of the scientific community and urged better means of incorporating it into collective decision making about managing the risks of emerging technologies. In a well-known paper about sheep farming in the northern United Kingdom after the Chernobyl nuclear accident, Wynne (1989) documented the "disconnect" between the knowledge of the radiation scientists brought in to evaluate safety issues and the knowledge of the sheep farmers concerning natural life cycles and other considerations arising outside of the scientific laboratory. The paper lucidly demonstrates that both forms of expertise, arising in radically different cultural settings, were valid and essential to appropriate risk management in this situation.

Wynne has been a leader in promoting public engagement efforts in the United Kingdom but recently resigned from a Food Standards Agency steering committee there, describing its efforts as largely intended to control publics on behalf of the biotechnology industry rather than to engage in sincere dialogue (GeneWatch "PR: New GM dialogue resignation welcomed." 2010). This is a reminder that while risk perception is cultural as well as scientific, it is most certainly also a political phenomenon that presents profound ethical challenges. Once again, risk communicators are reminded that (intentionally or not) their work sometimes serves particular interests (and value systems) over others, rather than simply providing disinterested scientific or other technical information to lay publics.

Acceptance or rejection of particular messages about risk relies heavily on whether people trust the sources of those messages. Although source credibility is an old story in communication research, the character of modern pluralistic democracies—in which people are blanketed with messages from a variety of sources, ranging from trusted leaders to sometimes mistrusted but often persuasive advertisers, from apparently disinterested scientists to advocates for or against particular scientific agendas—makes credibility and trust central issues for risk societies. Today, in an emerging trend, many companies try to "greenwash" themselves, painting their products and services with an aura of environmental respectability, sometimes justified and sometimes not. A handful of investment companies offer investment opportunities to concerned clients that the companies claim are tied to social justice considerations, a promising trend but also another claim that may or may not be true. Wise consumers use trust to negotiate through this landscape.

Trust is a very complex idea, however. For example, trust in environmental, scientific, and political or religious leadership appear to be independent constructs (Priest 2008). We may trust those we believe share our values, those we believe have special professional knowledge or expertise that we can rely on, or those we believe can be counted on to tell the truth (Barber 1983). A number of scholars around the world are currently investigating the nature of trust and the relationship between trust and message credibility. Generally speaking, however, conventional wisdom among communication specialists is that trust is likely to be both hard to gain and easy to lose, and then very difficult to regain. When sources of information become tainted by being associated with attempts to mislead or deceive, this is extremely difficult to repair. Risk communicators need to understand these dynamics and to try to persuade those they represent that honesty and transparency are the best policy for developing long-term credibility among those from whom they seek support, including consumers.

Persuasion Research

An extensive tradition within communication research concerns the nature of persuasion. This stream of research originated largely during World War II, when the U.S. federal government was trying to persuade people with money to invest to buy war bonds, housewives shopping for groceries to buy less desirable (and less familiar) cuts of meat in an era of shortage, and all citizens to support a war effort few fully comprehended. Many classic experiments date from this era. In especially well-known studies, researchers at Yale University investigated the effects of one-sided versus two-sided research with audiences of different education levels, concluding that two-sided messages—those that acknowledged both sides of an argument—were often more effective, especially with those of higher education who could be expected to understand that most all arguments have two sides (Hovland et al. 1949). Risk communication today follows much the same path, with enlightened risk communication specialists arguing that risks as well as benefits be acknowledged openly, regardless of who is disseminating these messages. Messages about emerging technologies that focus exclusively on benefits may be discounted.

In more recent years, much persuasion research within science communication has focused specifically on ways to communicate risk, appropriate levels of fear (given that too much fear causes people to shut down, and too little fails to motivate them to take action; Ruiter et al. 2001), and what factors ultimately motivate people to seek additional risk information on their own (risk information-seeking and processing theory; Griffin et al. 1999). Transparency and attention to two-way information transfer in risk communication are in the interests of everyone concerned. The purpose of risk communication for emerging technologies is not always clear, and this question has ethical dimensions. Is it for the purpose of manufacturing consent (Herman and Chomsky 1988) with respect to official positions—that is, for the purposes of persuading people to accept official or other expert definitions of which risks are worrisome and which are acceptable? Or is the ultimate purpose to empower individuals to make wise and informed risk decisions in their own lives, both as consumers and as citizens?

Nanotechnology risk communication can be analyzed as empowerment in a number of special situations (Priest 2009). Products containing nanotechnology may or may not carry information about this aspect on their labels, in the absence of law or regulation requiring this. Experience with GM foods strongly suggests that withholding information about product modification from consumers is simply a bad idea; consumer suspicion is inevitably aroused when it is later discovered that the product in question involves materials with special properties about which they were not made aware. The marketing of some early consumer products incorporating nanotechnology has been very open, using the nanotechnology element as a key selling point (as, for example, in the case of wrinkle-resistant clothing). Such products seem to have been readily embraced. When the arena moves to more sensitive areas such as medicine and food, consumer reaction may well be different, however.

As illustrated by the history of GM food, consumers generally feel they have a right to know about potentially risky technologies or materials to which they are being exposed, even for risks that appear minimal to experts. A strategy of nondisclosure can easily backfire. Future nanotechnology use in agriculture, such as the use of nanoparticles in pesticides or in the supermarket delivery of food products (e.g., in "smart" packaging that recognizes bacterial contamination), represents a potentially volatile situation. Medical applications are likely to be well received if they promise new treatment alternatives, but run some of the same risk of consumer rejection, especially if not transparently communicated. These biological (or "nanobiotechnology") applications arguably run the greatest risk of repeating the history of biotechnology, in terms of public perception.

Further, given the extensive promotion of nanotechnology as the source of new medical treatments and other benefits, issues of informed consent will inevitably arise as these treatments reach the public. More than likely, transparency is again going to be in everyone's best interests. Under the particular duress of confronting a potentially serious medical condition, it is highly desirable that citizens making decisions about medical treatment alternatives have as much prior background information as possible about what these alternatives, including those involving nanotechnology, might consist of. It is increasingly common for consumer and patient activism to accompany the development of new medical treatments for particular diseases. To the extent that nanotechnology promises to contribute to these new treatments, it is likely that this advocacy will be on the side of the technology's promoters. But unless communication about risks as well as benefits is open and transparent, ultimate consumer, citizen, and policy-maker reaction will remain unpredictable.

PERSPECTIVE: NANOTECHNOLOGY GOVERNANCE AND PUBLICS: MAKING CONNECTIONS IN POLICY

Jennifer Kuzma

Each nation, region, and state approaches policy design and implementation in different ways. For technology policy making in the United States and elsewhere, there has been a tendency to rely on technological "elites," such as industry developers, scientists and engineers, and government regulators, to determine and execute the rules for decision making. In other words, these elites are given the mandate of technological governance. There are windows of opportunity for broader stakeholder and public input, but they are small and often come late in the research, development, and marketing continuum. Where does the broader public have a choice about whether a technology is accepted or rejected in society? Where should it have say? This brief essay overviews key elements of public policy making in the United States as it relates to technological development, describes the current and potential roles of stakeholders and the broader public, and suggests ways in which the process could be opened to potentially lead to better and more democratic decisions.

Governance in its most basic sense is described as a "pattern of rule or activity of ruling" (Bevir 2009, 4). New governance models suggest roles for not only the state and markets but also stakeholders and citizens in the policy process. Some include the idea that networks of partnerships are superior to traditional state-run, top-down services or neoliberal market-based approaches to governance. In concert, social constructivism has arisen as a key concept in the field of science and technology studies (STS). STS scholars argue that because science and technology are socially constructed by the actors and networks involved—and thus reflect social values—participation on the part of the public in policy and decision making about them are warranted in a democracy. Ideas of new governance and social construction have sparked scholarly work around public engagement and participation in technological decision making, whereby social learning and communication are parts of a larger goal of giving people choice and voice. Under this framework, where does this occur currently, and where could it occur in the future for emerging technologies, such as bio- and nanotechnology?

One of the first decision points in technological development involves funding choices. The linear model of research and development (R&D) espoused by Vannevar Bush and others after World War II suggested that basic research funding would eventually lead to beneficial technologies for social good. However, STS and science and technology policy scholars have challenged this notion (e.g., Sarewitz 1996). Although there is wide recognition that some basic research can lead to applied science useful for solving environmental, public health, and social problems, it is not a given that the applications developed into products and processes will be directed by the market toward finding solutions to these problems. A balance needs to be struck between ensuring enough bottom-up basic research ideas from which new ideas and applications can freely arise, and basic and supporting applied research that is purposefully directed toward addressing societal problems. A new governance paradigm would suggest a need for public engagement at the beginning of S&T development, the earliest phases of funding of science and technology.

Currently, in the United States, these early stages of research investment are driven by the U.S. government or private industry. There is little to no place for public engagement in the *setting of research agendas.* Technical elites and interest groups may lobby for increased funding for certain areas such as defense, specific diseases, energy exploration, and consumer products, but mechanisms to open governance to the broader public and stakeholders are minimal to nonexistent. For example, the National Nanotechnology Initiative (NNI) in the United States compiles the funding for nanotechnology research and development. Budgets are set by the federal agencies in consultation with experts from industry and academe. There is little communication to the public about priority areas and the goals of U.S. funding. Some argue that the broader public is willing to accept decisions made by experts and agencies, or in other words, to give up this role in a representative democracy. However, given the interest of public participants in having a voice when educated about nanotechnology (Macoubrie 2005), a larger window of opportunity seems warranted. How and when to engage the broader public in helping to define R&D agendas remains a subject for research and analysis. In a recent window of opportunity, the NNI has invited comments on its new strategic plan (NNI 2010). Yet how these comments will be taken into consideration and whether the call for comments reached the interested layperson remain unknown.

One of the next steps in technological development is the hoped-for translation of basic and applied research into the development of socially useful products and applications. In private settings such as large corporations, this translation can occur in one institution. However, more often, there is a transfer of intellectual property from one institution to another, such as from federally funded academic or government laboratories to small biotechnology or nanotechnology companies or, in turn, from these small companies to larger companies. Translation comes through licensing of patents, material transfer agreements, and other business relationships such as consolidation or buyouts. The driving force comes from an entity that is willing to make the investment and take the risk to develop ideas and prototypes into products. Profit potential is the key motive. Government and other public-sector entities make few investments in technological development outside of military technology. As a result, most emerging technological products arise in areas with demonstrated money-making potential, such as pharmaceuticals for common diseases in developed countries (diabetes, heart disease, depression) or consumer products for widespread sale, also in developed countries.

Currently, the linear, market-driven model of technological development has led to thousands of consumer product applications of nanotechnology. The majority of these products are dietary supplements, sporting equipment, cosmetics, and personal care products. For example, there are very few products of nanotechnology on the market for environmental benefit (renewable energy, remediation) or addressing other societal problems like hunger or infectious disease. Under a new governance model, public consultation with government investment could conceivably drive emerging technologies more toward social good. This is a potentiality that has yet to be realized.

The broader public is generally more supportive of benefits to the environment and human health in emerging areas such as synthetic biology and nanotechnology than it is toward applications of market profit (e.g., Peter D. Hart Associates 2009). Given an opportunity to help decide what investments are made for the future of the planet, a broader set of stakeholders and publics could change the course of emerging technologies toward addressing societal problems. The system of innovation that includes public input would be consistent with evolving paradigms of new governance. At minimum, investments in emerging technologies should be made more transparent and communicated to the broader public. After all, the public is actually the initial investor via tax payments that go to initial stages of R&D.

Once a product is conceived and a prototype developed, companies must comply with oversight regimes, some of which involve mandatory systems of regulation. Much has been written on the strengths and weaknesses of U.S. regulation of nanotechnology products (e.g., Kuzma 2007; Davies 2009). Generally, most stakeholders, with the exception of corporations and companies promoting their own products, as well as the broader public have had little say in regulatory regimes and process. However, one window of opportunity for public input is through the *Federal Register.* Any citizen may submit comments through "notices of proposed rulemaking" on specific products or applications of technology (e.g., the Environmental Protection Agency's recent decision to regulate sliver nanoparticles under the Federal Insecticide, Fungicide, and Rodenticide Act). Although agencies are required to consider comments, how they consider them is up to their discretion. This window of opportunity for public input in regulation is fairly small and, in practice, mainly insignificant. There have been few studies of the impact of public comments on ultimate rule making for emerging technological products.

The broader public has little say in what technological products reach the market and the level of risk to which they (as consumers) are subjected. Nanoproducts do not need to be identified or labeled as such, according to current U.S. oversight policy. Furthermore, general policies to treat nanomaterials as substantially similar to their larger counterparts (e.g., the Food and Drug Administration's NanoTask Force report in 2007) are not subject to comment and rule-making, and a broad set of stakeholders and interested citizens are seldom invited to the table during the formulation of such policies. Again, this leaves users of such products with little to no choice to express their preferences on accepting nanotechnology (or not) in the items they use or consume.

Scholars have called for more deliberative democratic approaches to technological decision making all along the technological development continuum—from the creation of funding portfolios, to investments for product development, to the emergence of regulatory processes and approval, and to choice about consumption and use (e.g., Kuzma and Meghani 2009). Neoliberal approaches to technology governance, through markets and privatization, have not served society. Although technology has improved lives in some areas (e.g., longevity in developed countries, enhanced communication through the Internet), humanity is faced with perhaps the greatest problems ever in modern times. Threats to privacy, misuse of technology by terrorists, environmental degradation and pollution, natural resource overuse, global warming, growing inequities between the rich and poor, and infectious diseases in lesser developed countries can either stem from emerging technologies or be addressed by them. A new governance paradigm is being called for by experts, thinkers, scholars, and citizens to better guide policy on emerging technologies, like nanotechnology, toward addressing these global challenges. In the prophetic words of President Eisenhower in his January 1961 farewell address, "Yet, in holding scientific research and discovery in respect, as we should, we must also be alert to the equal and opposite danger that public policy could itself become the captive of a scientific-technological elite" (Eisenhower 1961).

Many who study the role of technology in society agree that change is warranted, although it is not clear how to instigate or implement a paradigm shift in this area. The current trend toward less government intervention in society will make a shift difficult, at least initially, until strong multi-institutional systems and collaborations are established to actively engage stakeholders and citizens in dialogue and decision making. Even so, questions of substance and process will remain. Scholars are just beginning to address the criteria and elements of effective deliberative, democratic technology-related decision making. Two articles of particular note are discussed below.

Bozeman and Sarewitz (2005) propose a "public failure model" to encompass a range of public values that can lead to failure of science and technology from the perspective of creating public good (not limited to "goods"). They propose criteria by which to judge whether public failures are likely to occur, much like "market failure models" are meant to assess how technologies may fail in the market if economic criteria are not considered. Public failure can occur when "when expression of public values is stifled or distorted" (123). Absence of mechanisms for nonexperts to have a voice in public investments and priorities related to science and technology is a primary cause of public market failures. Other causes occur when time horizons for decision making are too short so as to cause actions to eventually contradict public values and when benefits are inequitably distributed and captured by a select few as a result of the actions. The authors' criteria are designed to help judge whether science and technology decisions are compromising or are congruent with public values, regardless of the financial success of products or funding investments.

Also arguing for better mechanisms for public inclusion in science and technology policy systems, Jasanoff (2003) calls for "technologies of humility" that are "methods, or better yet institutionalized habits of thought" that "require not only the formal mechanisms of participation," like comment and rule making discussed above, but also "an intellectual environment in which citizens are encouraged to bring their knowledge and skills to bear on the resolution of common problems" (227). She suggests four elements through which to center the new technologies of humility: framing, vulnerability, distribution, and learning. These focus on questions of "what is the purpose; who will be hurt; who benefits; and how can we know?" (2003, 240). Process and substance, as well as deliberation and analysis, are equally important under the umbrella of these social "technologies of humility." Jasanoff states, "These approaches to decision-making would seek to integrate the 'can do' orientation of science and engineering with the 'should do' questions of ethical and political analysis" (2003, 243).

There are several points in the technological development continuum at which Bozeman and Sarewitz's "public failure model" and Jasanoff's "technologies of humility" could be deployed, as this article has discussed. What is currently missing is the momentum and political will to make the changes. Nanotechnology research and development could, however, serve as a test bed for these ideas among interested citizens, enlightened corporations, stakeholders, and scholars, until the rest of society catches up to embrace new governance paradigms.

REFERENCES

Bevir, M. 2009. *Key Concepts in Governance.* London: Sage.

Bozeman, B., and D. Sarewitz. 2005. Public values and public failure in U.S. science policy. *Science and Public Policy* 32(2): 119–136.

Davies, J.C. 2009. *Oversight for the Next Generation Nanotechnology.* Washington, DC: Project on Emerging Nanotechnologies.

Eisenhower, D.D. 1961. Military-Industrial Complex Speech. Public Papers of the Presidents, Dwight D. Eisenhower, 1960, 1035–1040. Available at http://www.h-net-org/~hst306/documents/indust.html.

Jasanoff, S. 2003. Technologies of humility: citizen participation in governing science. *Minerva* 41(3): 223–244.

Kuzma, J. 2007. Moving forward responsibly: oversight for the nanotechnology-biology interface. *Journal of Nanoparticle Research* 9: 165–182.

Kuzma, J., and Z. Meghani. 2009. A possible change in the U.S. risk-based decision making for emerging technological products: compromised or enhanced objectivity? *EMBO Reports* 10: 1–6.
Macoubrie, J. 2005. *Informed Public Perceptions of Nanotechnology and Trust in Government. The Project on Emerging Nanotechnologies.* Washington, DC: Woodrow Wilson International Center for Scholars and the Pew Charitable Trusts.
NNI (U.S. National Nanotechnology Initiative). 2010. "The National Nanotechnology Initiative Strategy Portal." Retrieved January 28, 2011, from http://strategy.nano.gov/.
Peter D. Hart Research Associates. 2009. Nanotechnology, synthetic biology, and public opinion. Washington, DC: Project on Emerging Nanotechnologies.
Project on Emerging Nanotechnologies. 2011. *Consumer Products Inventory.* Retrieved January 28, 2011. www.nanotechproject.org/inventories/consumer/.
Sarewitz, D. 1996. *Frontiers of Illusion*. Philadelphia: Temple University Press.

PERSPECTIVE: PUBLIC POLICIES AND PUBLIC PERCEPTIONS OF NANOTECHNOLOGY IN THE UNITED STATES AND IN EUROPE

Esther Ruiz Ben

Nanotechnology represents an emerging interdisciplinary technological and scientific method, which bears enormous expectations regarding its potential to fundamentally change the world. Nanotechnology could even bring the "Next Industrial Revolution" from the perspective of some policy advisors and technology developers (Schummer 2004). The development and application of this emerging method are, however, in an early state.

Some authors and institutions in different countries and regions have tried to develop particular definitions of nanotechnology in order to build a public understanding about this innovation (e.g., the report for the technology assessment group of the German Parliament: Paschen et al. 2003; the British Standards Office's definitions PAS71 2005). However, such definitions and the media coverage of nanotechnology differ among countries and are linked to different policies and social as well as cultural and economic aspects. Accordingly, the public perception of nanotechnology differs very much among countries as some studies have revealed (e.g., Gaskell et al. 2005; Kjærgaard 2010).

In general terms, some authors have pointed out that a possible analogy between nanotechnology and genetically modified food could arise, bringing a negative public stigmatization (Wilkinson et al. 2007) of nanotechnology (e.g., Friedman and Egolf 2005; Pidgeon and Rogers-Hayden 2007; Throne-Holst et al. 2007). Because of this threat of stigmatization of nanotechnology, some authors have called for an active engagement of the general public in a dialogue with policy makers and developers about the development of nanotechnology, which has been labeled *upstream* engagement (Pidgeon and Rogers-Hayden 2007). How is such a dialogue constructed in regions with different cultural, social, and economic trajectories? What are the perceptions of nanotechnology in different national and regional areas? In this brief chapter based upon secondary literature, I concentrate on the public policies and public perceptions of nanotechnology in two geographic areas that are widely contributing to the development of this innovation in the world: the United States and Europe (Palmberg et al. 2009).

PUBLIC POLICY OF NANOTECHNOLOGY IN THE UNITED STATES AND EUROPE

In the United States, media coverage of nanotechnology has been dominated in past years by optimistic science and business writers. It has focused on social benefits and economic development with the result that those persons who were aware of nanotechnology were enthusiastic about it, as Nisbet and Scheufele (2007) argue.

A relevant example of the concept of such optimistic public policy regarding nanotechnology can be found in an article by Roco and Bainbridge (2004) which summarizes opinions presented in a workshop organized by U.S. National Nanotechnology Initiative (NNI) in 2003. In this summary the public represents a target group for actions but not an active group. The goal is to achieve an "informed population" in order to prevent negative attitudes against nanotechnology.

Also, a more recent analysis of print media (570 print media articles between 1998 and 2005) conducted by Fitzgerald (2006) revealed the more prevalent coverage of positive aspects of nanotechnology than risks. The most frequently mentioned benefits were related to the enhanced quality of goods and services as well as prevention of diseases, whereas the most frequently cited risks referred to health problems and environmental issues.

In the European context, public policies have been very diverse. Some countries have conceptualized users of nanotechnology as citizens seeking engagement in nanotechnology policy, which is the case in the United Kingdom, but in other countries like Denmark or Germany, such practices are not known. Concretely, in the United Kingdom, public policy regarding nanotechnology has followed a citizens' jury method. The NanoJury U.K. was organized in 2005 and generated several recommendations regarding issues such as the public funding of nanotechnology development and the models for public engagement (Pidgeon and Rogers-Hayden 2007).

Unfortunately, the research about public policy practices regarding nanotechnology has concentrated on the situation in the United States, mostly analyzing media coverage (Friedman and Egolf 2005; Gorss and Lewenstein 2005; Stephens 2005; Faber 2006) and in the United Kingdom (Anderson et al. 2005; Pidgeon and Rogers-Hayden 2007). Some exceptions are Denmark (Kjærgaard 2010) and the Netherlands. Te Kulve (2006) argues, based on an analysis of the Dutch newspaper reports about nanotechnology, that from the beginning of media coverage on this issue, both high and more modest expectations were present. Particularly in the years 2001 and 2003 in the Netherlands and in the United Kingdom, new projects about nanotechnology were founded, particularly in medicine, biotechnology, and information technology (IT) (Te Kulve 2006, 374–376). Nevertheless, even though a Micro and Nanotechnology Network (MNT Network) was established in 2003 with the idea of supporting the industry in the development of commercial potentials of nanotechnology, this issue has received little attention in the British media (Anderson et al. 2005).

From a comparative perspective in the United States, British, Dutch, and Danish newspapers, Kjærgaard (2010) analyzed the rise and fall in newspaper coverage of nanotechnology, as well as the development of particular news frames using the model of a "circuit of mass communication." Kjærgaard concludes that the Danish and the Dutch cases show a complex picture within the European context. National differences and local conditions appear to be more influential than geographic proximity and demographic similarity. But even if the analyses of media coverage clearly reflect the contents and tones of the political agenda of these countries, they do not say much about public understanding and public opinion of nanotechnology. The relationship between media and the public is highly complex, and the question of what influences public opinion and to what extent is difficult to answer.

PUBLIC PERCEPTION OF NANOTECHNOLOGY IN THE UNITED STATES AND IN EUROPE

In studies of U.S. and British media coverage of nanotechnology, a prominence of the "science fiction and popular culture" frame has been found. Several scholars refer to a *Prey effect* in the United States following Michael Crichton's best-seller publication in 2002 which portrays the threatening scenario of autonomous nanomachines (Gorss and Lewenstein 2005; Bowman and Fitzharris 2007; Kjærgaard 2010, 90). Cobb and Macoubrie (2004) initially examined, on the basis of a survey with 1536 randomly selected persons in the United States, the association of science fiction writing and the public's perception of nanotechnology. They found an influence of Crichton`s book in the respondents that afterwards was called the *Prey* effect, noting that "survey data show that public opinion is negatively affected by knowing the details of *Prey*" (404). In a further study, however, Cobb (2005, 235) found that many in the United States have a positive view of nanotechnology that remains positive even after being exposed to negative frames. Thus, the *Prey* effect may be quite relative. The particular dynamics of this effect among science fiction readers—those most likely to have read *Prey*—should be taken into account.

In the United Kingdom, the influence of "science fiction and popular culture" is related to some public comments of Prince Charles on nanotechnology (Anderson et al. 2005, 216). Kjærgaard concludes in his analysis that *no comparable examples are found in either the Dutch or the Danish contexts, marking a difference within European countries* (2010, 87). This study contradicts the conclusions of Gaskell et al. (2005) who compared in a previous study U.S. and European attitudes toward nanotechnology. In general, they found much public uncertainty about nanotechnology, but the U.S. subjects were on the whole significantly more optimistic, as they were regarding other technologies. As a conclusion, they suggest that the attitudes toward nanotechnology are more positive in the United States than in Europe:

> We suggest that people in the U.S. assimilate nanotechnology within a set of pro-technology cultural values. By contrast, in Europe there is more concern about the impact of technology on the environment, less commitment to economic progress and less confidence in regulation. These differences in values are reflected in media coverage, with more emphasis on the potential benefits of nanotechnology in the U.S. than in the U.K.

> Finally, we speculate on possible futures for the reception of nanotechnology in the U.S. and Europe.

Kjærgaard (2010, 87) demonstrates that the variations in the specific regional or national contexts are at least equally important in shaping the local narratives dominating the framing of nanotechnology. This coincides with the results of the analysis conducted by Bauer (2009):

> The survey evidence shows that the public understanding of science might be significantly different in an industrial-developing context and a knowledge-intensive developed context. In the latter more knowledge does not bring more support for science, rather utilitarian scrutiny, and an end to widespread beliefs in ideology and myths of what science might be.

One common result of several studies about the knowledge of nanotechnology in some European countries and in North America is that the public knew little about this innovation (for the United Kingdom, this comes from the Royal Society and Royal Academy of Engineering 2004; for the United States and Canada, from Priest 2006; for Japan, from the Nanotechnology Research Institute 2006; for Germany, from Vandermoere et al. 2010; and for France, from Vandermoere et al. 2009). However, the question of the relationship between information and awareness about nanotechnology and the attitudes toward it is not easy to answer.

Although some studies have tried to emphasize the positive relationship between information and attitudes toward nanotechnology and some have even found a corresponding relationship (Satterfield et al. 2009) based on meta-analysis, other research analysis has demonstrated that this relationship can be quite limited (Sturgis and Allum 2004; Bauer et al. 2007). Other scholars have found, through the use of more control variables in their analysis, that general basic science literacy is a better predictor of positive attitudes toward nanotechnology than the more concrete awareness about nanotechnology (Lee et al. 2005; Scheufele and Lewenstein 2005; Cacciatore et al. 2009). An international comparison addressing the influence of those variables in different national contexts does not yet exist.

REFERENCES

Anderson, A., S. Allan, A. Petersen, and C. Wilkinson. 2005. The framing of nanotechnologies in the British Newspaper Press. *Science Communication* 27(2): 200–220.

Bauer, M.W. 2009. The evolution of public understanding of science—discourse and comparative evidence. *Science, Technology and Society* 14(2): 221–239.

Bauer, M., N. Allum, and S. Miller. 2007. What can we learn from 25 years of PUS survey research? Liberating and expanding the agenda. *Public Understanding of Science* 16(1): 79–95.

Bowman, D.M., and M. Fitzharris. 2007. Too small for concern? Public health and nanotechnology. *Australian and New Zealand Journal of Public Health* 31(4): 382–384.

Cacciatore, M.A., D.A. Scheufele, and E.A. Corley. 2009. From enabling technology to applications: the evolution of risk perceptions about nanotechnology. *Public Understanding of*Science. doi: 10.1177/0963662509347815.

Cobb, M.D. 2005. Framing effects on public opinion about nanotechnology. *Science Communication* 27(2): 221–239.

Cobb, M.D., and J. Macoubrie. 2004. Public perceptions about nanotechnology: risks, benefits and trust. *Journal of Nanoparticle Research* 6(4): 395–405.

Faber, B. 2006. Popularizing nanoscience: the public rhetoric of nanotechnology 1986–1999. *Technical Communication Quarterly* 15(2): 141–169.

Fitzgerald, S.T. 2006. Constructing risk: media coverage of nanotechnology. Paper presented at the annual meeting of the American Sociological Association, Montreal, Quebec, Canada. http:www.allacademic.com//meta/p_mls_apa_research_citation/1/0/4/6/8/pages104680/p104680-1.php.

Friedman, S.M., and B.P. Egolf. 2005. Nanotechnology: risk and the media. *IEEE Technology and Society Magazine* 4: 5–11.

Gaskell, G., T. Ten Eyck, J. Jackson, and G. Veltri. 2005. Imagining nanotechnology: cultural support for technological innovation in Europe and the United States. *Public Understanding of Science* 14(1): 81–90.

Gorss, J., and B.V. Lewenstein. 2005. The salience of small: nanotechnology coverage in the American press, 1986–2004. Paper presented at the *Annual Conference of the International Communication Association*, 26–30 May, New York.

Kjærgaard, R.S. 2010. Making a small country count: nanotechnology in Danish newspapers from 1996 to 2006. *Public Understanding of Science* 19(1): 80–97.

Lee, C.J., D.A. Scheufele, and B.V. Lewenstein. 2005. Public attitudes toward emerging technologies: examining the interactive effects of cognitions and affect on public support for nanotechnology. *Science Communication* 27(2): 240–267.

Nanotechnology and Society Survey Project. 2006. Perception of nanotechnology among general public in Japan. Nanotechnology Research Institute. Retrieved from www.nanoworld.jp/apnw/articles/library4/pdf/4-6.pdf.

National Nanotechnology Initiative. History. 2010. Washington, DC: National Science Foundation (updated December 2010). Retrieved from www.nano.gov/html/about/history.html.

Nisbet, M.C., and D.A. Scheufele. 2009. What's next for science communication? Promising directions and lingering distractions. *American Journal of Botany* 96(10): 1767–1778.

Palmberg, C., H. Dernis, and C. Miguet. 2009. Nanotechnology: an overview based on indicators and statistics. OECD: Paris: *STI Working Paper* 2009/7. JT03267289.

PAS71. 2005. Publicly available specification vocabulary—nanoparticles. London: British Standards Institution.

Paschen, H., C. Coenen, T. Fleischer, R. Grünwald, D. Oertel, and C. Revermann. 2003. Nanotechnologie: Forschung, Entwicklung, Anwendung. Berlin: Springer (TAB-Arbeitsbericht 92, Berlin: Büro für Technikfolgen-Abschätzung beim Deutschen Bundestag).

Pidgeon, N., and T. Rogers-Hayden. 2007. Opening up nanotechnology dialogue with the publics: risk communication or 'upstream engagement.' *Health, Risk and Society* 9(2): 191–210.

Priest, S. 2006. The North American opinion climate for nanotechnology and its products: opportunities and challenges. *Journal of Nanoparticle Research* 8: 563–568.

Roco, M.C., and W.S. Bainbridge. 2004. Societal implications of nanoscience and nanotechnology: maximizing human benefit. *Journal of Nanoparticle Research*. 7: 1–13.

Royal Society and The Royal Academy of Engineering. 2004. *Nanoscience and Nanotechnologies: Opportunities and Uncertainties*. London.

Satterfield, T., M. Kandlikar, C.E.H. Beaudrie, J. Conti, J. Herr, and B. Harthorn. 2009. Anticipating the perceived risk of nanotechnologies. *Nature Nanotechnology* 4(11): 752–758.

Scheufele, D.A., and B.V. Lewenstein. 2005. The public and nanotechology: how citizens make sense of emerging technologies. *Journal of Nanoparticle Research* 7: 659–667.

Schummer, J. 2004. Societal and ethical implications of nanotechnology: meanings, interest groups and social dynamics. *Techné* 8(2): 56–87.

Stephens, L.F. 2005. News narratives about nano S&T in major U.S. and non-U.S. newspapers. *Science Communication* 27(2): 175–199.

Sturgis, P., and N. Allum. 2004. Science in society: re-evaluating the deficit model of public attitudes. *Public Understanding of Science* 13(1): 55–74.

Te Kulve, H. 2006. Evolving repertoires: nanotechnology in daily newspapers in the Netherlands. *Science as Culture* 15(4): 367–382.

Throne-Holst, H., E. Stø, and U. Kjærnes. 2007. Governance issues for emerging technologies: The GM food—NANO discourses. *Proceedings of the Nordic Consumer Policy Research Conference* 2007, 1–14.

Vandermoere, F., S. Blanchemanche, A. Bieberstein, S. Marette, and J. Roosen. 2010. The morality of attitudes toward nanotechnology: about God, techno-scientific progress, and interfering with nature. *Journal of Nanoparticle Research* 12(2): 373–381.

Vandermoere, F., S. Blanchemanche, A. Bieberstein, S. Marette, and J. Roosen. 2009. The public understanding of nanotechnology in the food domain: the hidden role of views on science, technology, and nature. *Public Understanding of Science.*

Wilkinson, C., S. Allan, A. Anderson, and A. Petersen. 2007. From uncertainty to risk?: Scientific and news media portrayals of nanoparticle safety. *Health, Risk and Society* 9(2): 145–157.

References

American Cancer Society. 2010. Cancer facts and figures. Atlanta, GA: American Cancer Society. Retrieved January 10, 2011 from www.cancer.org/acs/groups/content/@epidemiologysurveilance/documents/document/acspc-026238.pdf.

Barber, B. 1983. *The Logic and Limits of Trust.* New Brunswick, NJ: Rutgers University Press.

Berger, P.L., and T. Luckmann. 1991. *The Social Construction of Reality: A Treatise in the Sociology of Knowledge.* London: Penguin Books.

Bradbury, J.A. 1989. The policy implications of differing concepts of risk. *Science, Technology and Human Values* 14(4): 380–399.

Cobb, M., and J. Macoubrie. 2004. Public perceptions about nanotechnology: risks, benefits and trust. *Journal of Nanoparticle Research* 5: 395–405.

Cole, L.W., and S.R. Foster. 2001. *From the Ground Up: Environmental Racism and the Rise of the Environmental Justice Movement.* New York: New York University Press.

Douglas, M., and A. Wildavsky. 1983. *Risk and Culture: An Essay on the Selection of Technical and Environmental dangers.* Berkeley: University of California Press.

GeneWatch "PR: New GM dialogue resignation welcomed." 2010, May. *GeneWatch UK.* Retrieved January 9, 2011, from www.genewatch.org/article.shtml?als[cid]=405258&als[itemid]=566339.

Griffin, R.J., S. Dunwoody, and K. Neuwirth. 1999. Proposed model of the relationship of risk information seeking and processing to the development of preventive behaviors. *Environmental Research* 80(2): S230–S245.

Herman, E.S., and N. Chomsky. 1988. *Manufacturing Consent: The Political Economy of the Mass Media.* New York: Pantheon Books.

Hovland, C.I., A.A. Lumsdaine, and F.D. Sheffield. 1949. *Experiments on Mass Communication.* Princeton: Princeton University Press.

Jordan, W. 2010, (April 29). Nanotechnology and Pesticides. Presentation to the Pesticide Programs Dialogue Committee. Washington, DC. Retrieved January 12, 2011, from www.nanotechproject.org/process/assets/files/8309/epa_newpolicy_nanomaterials.pdf.

Kuzma, J., and S. Priest. 2010. Nanotechnology, risk, and oversight: Learning lessons from related emerging technologies. *Risk Analysis* 30(11): 1688–1698.

Maynard, A. 2008. "U.S. nanotechnology risk research funding—separating fact from fiction." Retrieved January 10, 2011, from http://community.safenano.org/blogs/andrew_maynard/archive/2008/04/18/u-s-nanotechnology-risk-research-funding-separating-fact-from-fiction.aspx.

National Institute for Occupational Safety and Health. 2009. Approaches to safe nanotechnology: managing the health and safety concerns associated with engineered nanomaterials, NIOSH Publication No. 2009-125. Retrieved January 10, 2011, from www.cdc.gov/niosh/docs/2009-125/pdfs/2009-125.pdf.

National Science and Technology Council Committee on Technology, Subcommittee on Nanoscale Science, Engineering and Technologies. 2010. *The National Nanotechnology Initiative: Research and Development Leading to a Revolution in Technology and Industry,* supplement to the president's 2011 budget in Washington, DC: www.nano.gov/NNI_2011_budget_supplement.pdf (accessed January 9, 2011).

Pardo, R., and F. Calvo. 2008. Attitudes toward embryo research, worldviews, and the moral status of the embryo frame. *Science Communication* 30(1): 8–47.

Poland, C.A., R. Duffin, I. Kinloch, A. Maynard, W.A.H. Wallace, A. Seaton, V. Stone, S. Brown, W. MacNee, and K. Donaldson. 2008. Carbon nanotubes introduced into the abdominal cavity of mice show asbestos-like pathogenicity in a pilot study. *Nature Nanotechnology* 3(7): 423–428.

Priest, S.H. 2008. North American audiences for news of emerging technologies: Canadian and U.S. responses to bio- and nanotechnologies. *Journal of Risk Research* 11(7): 877–889.

Priest, S. 2009. Risk communication for nanobiotechnology: to whom, about what, and why? Special issue on "Developing Oversight Approaches to Nanobiotechnology: The Lessons of History," *Journal of Law, Medicine, and Ethics* 37: 759–769.

Priest, S., and T. Greenhalgh. n.d. Nanotechnology as an experiment in democracy: how do citizens form opinions about technology and policy? *Journal of Nanotechnology Research*. Paper (to be) presented at the Society for Nanotechnology and Emerging Technologies Annual Meeting, Arizona State University, October 2011.

Priest, S., T. Lane, T. Greenhalgh, L.J. Hand, and V. Kramer. In press. Envisioning emerging nanotechnologies: a three-year panel study of South Carolina citizens. *Risk Analysis*. Unpublished paper.

Ruiter, R.A.C., C. Abraham, and G. Kok. 2001. Scary warnings and rational precautions: a review of the psychology of fear appeals. *Psychology and Health* 16(6): 613–630.

Siegrist, M. 2010. Psychometric paradigm. In *Encyclopedia of Science and Technology Communication* 2. Ed. S.H. Priest. Thousand Oaks, CA: Sage Publications: 600–601.

Slovic, P. (2000). *Perception of Risk*. London: Earthscan.

Slovic, P., B. Fischhoff, and S. Lichtenstein. 1981. Perceived risk: psychological factors and social implications. In *The Assessment and Perception of Risk*. Eds. F. Warner and D.H. Slater. London: The Royal Society, 17–34.
Song, Y., X. Li, and X. Du. 2009. Exposure to nanoparticles is related to pleural effusion, pulmonary fibrosis and granuloma. *European Respiratory Journal* 34(3): 559.
Tversky, A., and D. Kahneman. 1986. Rational choice and the framing of decisions. *Journal of Business* 59(4): 251–278.
Wynne, B. 1989. Sheep farming after Chernobyl: a case study in communicating scientific information. *Environment: Science and Policy for Sustainable Development* 31(2): 10–39.

7

Disseminating Information about New Technologies

The movement toward *upstream* public engagement in making decisions about emerging technology, including nanotechnology, is a positive development that reflects optimism about the possibility of practicing better democracy in a risk society. However, not everyone is going to be reached by such programs, and their direct influence is thus quite limited at this stage. The challenges involved in organizing and funding even modest community engagement programs are formidable, and not all citizens will choose to participate in these opportunities (Kleinman et al. 2009; Powell 2008). The traditional way that risk communicators, like other communication specialists interested in reaching broader publics, get their messages disseminated is primarily through the mass media, using the tried-and-true method of issuing press releases to alert journalists to potential stories (but with one important change in that press releases today are almost always electronic, whether distributed by a wire service or sent directly to media organizations or editors). Under some circumstances, public service announcements and even commercial advertising also play roles. Advertising rarely focuses on risks unless this is required (e.g., in the United States, by the U.S. Food and Drug Administration, FDA), and it raises awareness of products and may result indirectly in incidental learning about risks. But the goal of distributing press releases is influencing the news.

The media not only serve as the main information source for most adults about emerging new technologies (as about most other contemporary news issues), they also provide news about public opinion and political debate—potentially, even news about engagement activities. In so doing, they are capable of acting as a magnifier for the influence of those activities; for example, the news media routinely alert us to political debates that would otherwise only influence those directly involved, and news stories about engagement activities could have the same type of effect. Even those who are not participating in discussions on contemporary issues can get information, be exposed to arguments, gain some sense of the climate of public opinion, and form their own ideas based on media discourse. In an ideal world, information from the mass media would support productive *public sphere* discussion by providing alternative views, as well as news, about emerging technologies as about all other areas of public life. The news media can thus provide a sort of vicarious opportunity for engagement; even individual citizens acting alone can compare their opinions and reactions to those of others, assess available arguments and positions, and make up their minds—somewhat as though they were actually engaged in "live" discussion and debate.

Of course, the reality generally falls short of this ideal; both structural and economic issues generally prevent the mass media from fully realizing this potential. Most modern news media organizations operate as commercial entities and must seek audiences and advertisers. Even nonprofit or government-supported media (such as NPR or PBS in the United States or the BBC in the United Kingdom) compete with for-profit media organizations in these respects. Stories about noncontroversial new technologies are not generally leading news items. News of political debate is often confined to public service channels or other specialized media. Complex scientific and technical developments are not easy to turn into readable and entertaining stories (see, e.g., Friedman et al. 1986). Scientists, even those who are very concerned with scientific knowledge among members of nonexpert publics, may decline to talk to journalists because they are afraid of being misquoted or otherwise misrepresented, especially if the topic is a controversial one involving such things as environmental or health risks. Journalists are not always anxious to work with such reluctant sources. Editors are increasingly reluctant—especially in tough economic times—to assign journalists to complex stories requiring extensive research and steep learning curves. Journalists who specialize in science and technology are becoming a rarity, perceived by some producers and editors as a dispensable luxury.

As a result, although the conventional news media are a key vehicle through which the *marketplace of ideas* must operate, media often report science stories on a limited basis at most. Perhaps more accurate, on average, than many scientists and other technical specialists are inclined to assume, some news coverage of science and technology is superficial; other coverage focuses narrowly on allegedly "breakthrough" research, individual studies promoted by enterprising public information officers at specific universities and research institutes, rather than attempting to capture the nature of emerging scientific consensus involving trends across many studies. Journalists, even science journalists, generally do not see their role as "educating the public," but as reporting on the vital news of the day; editors do not want to "spend" precious space reporting on these long, complex stories, especially those that they do not believe (rightly or wrongly) will captivate audiences. Essential background information may therefore be missing from science in the media, and without appropriate context, the result can be stories that are difficult for nonspecialist readers to understand.

Increasingly, the marketplace of ideas is shifting to the Internet, where information may be presented by traditional news organizations or by any number of other organizations. This solves some problems (such as severe limits on space) but creates others. The news media, however imperfect, generally have strong traditions concerning accuracy and balance (although nations also differ widely in the exact expectations associated with the appropriate, ethical practice of journalism). These variations aside, the job that all news media do for society is select, filter, and shape the news, even though just how they do this job has been the subject of extensive criticism, including the criticism that they routinely overrepresent the views of the powerful and ignore dissenting or minority opinion. Representatives of other organizations who are also seeking to promote ideas, disseminate information, or shape opinion are rarely under the same obligation to serve the general interests of society as the news media, but can it be equally influential, not only through public information and public relations activities but also by acting as key sources for journalism. Such entities include any of the stakeholder organizations discussed in previous chapters of this book, from research organizations to consumer advocacy groups, to promoters and marketers of specific commercial products, and from government agencies to political figures and lobbyists.

While in theory these dynamics might produce a wide "marketplace" from which consumers can choose information and ideas, with the Internet facilitating its vitality, in practice information-seeking citizens are well advised to be wary consumers of all information, in particular, Internet-based information. Climate change is a good example of why this matters and how it can be expected to matter more in the future. Conventional media coverage has been criticized for presenting climate change—in conventional journalistic style—as a "two-sided" issue (Boykoff and Boykoff 2007), even though a strong scientific consensus supports the view that global climate change is occurring and is in all likelihood anthropogenic. This has been a fair criticism and one to which the mass media are beginning to respond. Yet it is easy to find Internet-based Web sites claiming that climate change is not taking place at all, often presenting what may appear to be "scientific" support for their counterclaims. Today's citizens will need new information-processing skills to guide them through a maze of claims and counterclaims on many issues. On the one hand, this phenomenon could promote healthy debate on some topics; on the other, it is going to be increasingly up to individual information consumers to sort out credible views from impossible ones, and to do so without the guideposts provided by traditional news organizations, however imperfect these may have been.

The future of the conventional news media is far from certain. Print journalism in the traditional sense is shrinking. A new business model that makes Internet news distribution commercially viable has been elusive. Broadcast news may not be far behind; the time constraints on conventional news programs are severe, and even broadcasters on a 24/7 news cycle face all the same economic and practical constraints on reporting news about science and technology as do the more traditional media organizations. What will the world of news and information be like when scientific and technological developments are presented primarily by public relations specialists operating Web sites on behalf of particular stakeholders? One thing is certain: Risk communication will take place, in the future, in a changing media world, even though we do not yet know exactly what changes will unfold. The challenge will be to define new guidelines for what constitutes responsible risk communication, when the traditional guidelines of journalistic practice may no longer apply.

Media's Role in Risk Societies

Sociologists look at the role of social institutions through a number of different theoretical lenses. One of these is that of "functionalist" sociology, which sees the structure of society as composed of a variety of different groups, organizations, and institutions that operate together more or less smoothly to meet society's needs. Another is conflict theory, which argues that, on the contrary, societies are best understood as composed of groups that compete for power, which is unequally distributed. In this view, mainstream social institutions often serve to support the existing status quo distribution of power; social change comes about only as other groups challenge this power, sometimes via very widespread social movements such as the environmental, civil rights, women's rights, and gay rights movements. From a functionalist point of view, inadequacies and inequities in meeting society's needs reflect a system that is merely out of adjustment, in which relatively minor changes can generally address problems successfully. From a conflict theoretic point of view, competition for power is an inherent characteristic of human social organization, and the assumption that the system will right itself absent deliberate social action is a problematic one at best. Both of these perspectives have been used to help illuminate the role of communication media, especially the news media, in society.

Functionalist sociologists often seek to describe the purposes, or functions, that different institutions serve for society. Media organizations have been described as having at least four basic social functions: surveillance of the environment (looking out for threats, a function of clear relevance to risk communicators), correlation (that is, providing context by relating different events to one another, something that also aids effective risk communication), transmission of culture or social heritage (teaching new generations about social values and expectations, which new risks may challenge), and entertainment (Lasswell 1948; Wright 1986). The inclusion of "entertainment" on this list might seem odd when we think in terms of news media specifically, in fact even news material must capture the interests and attention of news consumers to have any influence at all, and to provide a means to sell the advertising that has historically kept news organizations profitable, audiences of consumers are essential. This element, too, is relevant to risk communication, because without audience attention, risk-related messages have no impact, but with too much attention, risks can become exaggerated. With the advent of the Internet as a primary direct provider of news and information of all sorts directly to consumers, perhaps this model will change. But the need to attract and hold audiences is going to remain with us in any press system set up on a for-profit basis, whether the channel involved is print, broadcast, or the Internet.

Multiple consequences follow from thinking about the press as a set of businesses that fulfill certain functions or obligations on society's behalf but are also driven by the need to turn a profit. On the one hand, press accounts of technological risk are regularly accused of being overly sensationalistic in their attempts to draw attention to their content in order to attract and hold audiences. This observation goes a long way toward explaining the observation of social amplification theory (Kasperson et al. 2003; see also Chapter 3) that the media play a key role in amplifying risks. In practice, social amplification studies have more often been concerned about the amplification of risk (that is, cases where experts think a risk is being exaggerated in media accounts and elsewhere within society) than about the attenuation of risk (that is, cases where experts think a risk is being understated rather than overstated). Criticisms of media performance, even if not informed by this theory, often focus specifically on the tendency to be sensationalistic when reporting risks. "If it bleeds, it leads" is a familiar saying to broadcast journalists—a graphic and succinct statement of much the same point.

On the other hand, the press can also sometimes be accused of being complacent about risks, of ignoring conflict, or even of covering up bad consequences (e.g., poor government decision making or ineffective policies). If no dramatic outcomes such as death, disease, or visible environmental degradation can be observed, and especially if "watchdog" advocacy groups are silent, there is little internal incentive to cover the story. Although our stereotype of the practice of journalism may imagine journalists as enterprising righters of wrongs trying to ferret out important stories that might bring them an exclusive "scoop" on a breaking issue, the economics of the news business dictate that this is more often the exception rather than the rule. If the risks of nanotechnology are socially attenuated, as suggested earlier in this book, these same economics may be partly to blame (even though more general cultural factors may be yet more important). These dynamics pose a very special challenge for risk communicators who might want to call attention to risks without exaggerating them (e.g., in the case of nanotechnology, by inadvertently fostering a tendency to generalize from specific areas where concern seems warranted to all nanotechnology).

Fears of being featured in sensationalistic coverage likely underlie some of the reluctance of scientists to interact with the journalistic community. In turn, given these dynamics, the role of public relations and public information specialists becomes that much greater. News about science, technology, and risk is sometimes described as "subsidized" news (Gandy 1982), dependent on "information subsidies" from those handing out free information in the form of press releases or statements made at press conferences, whether industry, government, research institutions, or nonprofit groups. Those providing the information often get to shape the news that results, even if responsible journalists try not to give any single source too much weight. In an era in which little money is invested by news organizations in investigative work, given topics that are inherently complex and technical and in which there is little prior public interest, it is not particularly surprising that journalists are not always proactive in seeking risk-related information. "Booster" coverage that emphasizes benefits alone may be typical of subsidized news about emerging technology, but both society at large and technology's key stakeholders may be better served by a more "balanced" view. How balance will be maintained in the age of the Internet remains to be seen.

Conflict theory, functionalism's chief theoretical rival within sociology, proposes that society is best thought of not as composed of institutions that generally work smoothly together to meet society's needs, but as made up of groups with different economic and other interests that are often at odds with one another, in an environment in which power is unequally distributed. Karl Marx, observing the industrial revolution of the nineteenth century, originally conceptualized this as a matter of class conflict (1999). Modern versions of conflict theory recognize many divisions within society, including gender and ethnic/racial divisions as well as socioeconomic class divisions, as being related to unequal distribution of power, different economic interests, and unequal access to resources. Although functionalism and marketplace of ideas theory suggest that many voices should participate in public dialogue, voices that media strive to represent appropriately, conflict theory emphasizes that all voices are not equal—that, at a minimum, powerful interests are likely to have more access to the resources necessary to promote their points of view than less powerful interests. From this perspective, media are under implicit (or even explicit) pressure to reflect the views of powerful interests within society, and cannot be assumed to be responsive to all views equally. At a minimum, the resources available to more powerful groups to encourage media to represent their views create marketplace inequities.

Both functionalism and conflict theory are helpful for understanding how risk communication actually works. From a functionalist perspective, the media's role in reporting on risks is in large part to point them out ("surveillance of the environment") and report on society's responses to those risks ("correlation"). Ideally, this involves providing a marketplace of competing perspectives. In the process, the media convey, perhaps unintentionally, a sense of how we should think of both technology and its inevitable risks ("transmission" of culture). To do so, they must grab and hold our attention ("entertainment"), resulting in at least some degree of bias toward sensationalizing risk information and reporting most heavily on more extreme threats. However, conflict theory argues that media organizations are likely to be more responsive to the interests of some of the more powerful groups within society, not always to all interests equitably. Those with power generally include the promoters of particular technologies.

Similarly, risk communicators most often represent those with resources to employ them. Journalists tend to represent the perspectives of their sources. For nanotechnology, sources journalists utilize inevitably include the technology's promoters—the agencies, universities, and commercial concerns for which nanotechnology represents an important economic investment and a potential economic opportunity. This can and does generate "booster" coverage, especially in the early stages of a new technology's appearance on the public stage, which focuses on the positive hopes and may tend to avoid discussions of risk. But some risk communicators providing information to the journalistic community also represent nonprofit groups and other agencies and organizations whose interests include the active protection of consumers and citizens, as well as those promoters who have recognized that balanced coverage often works in their interests as well. From a functionalist perspective, this system may be seen to work well, producing a true marketplace of ideas, although perhaps one with a built-in bias toward sensationalizing (or "amplifying") risk. However, from a conflict-theoretic perspective, the journalistic landscape is more likely to be seen as remaining skewed toward serving more powerful groups, who may sometimes be more interested in risk attenuation than risk amplification. Certainly there is a grain of truth in both of these views.

The interests of promoters and watchdogs may converge in some important respects, however, and this is certainly important from a risk communication perspective. It is arguably in the overall interests of nanotechnology's promoters to recognize risks and to advocate for their proactive management by society, if only to avoid an otherwise inevitable negative public backlash. Conversely, it is in the interests of watchdog organizations to protect the potentially positive social contributions of technology on behalf of those they represent, as well as to protect society from potentially negative effects (that is, from risks). This is as true for nanotechnology as it is for any other set of emerging technologies. In any event, media inattention to the likely future effects of nanotechnology, both good and bad, would not seem to serve anyone's interests. Broad public discussion of both risks and benefits is needed. Ignoring the potential for risk, or even the potential for the *perception* of risk, does not serve the long-term interests of promoters, as the history of other new technologies clearly demonstrates. Public engagement in consideration of risks seems a vital step in technology adoption, and media discussion is a vital complement to increased public engagement.

Given the common criticisms of media coverage of technological risk, sometimes society at large seems to expect that journalists will be capable of being dispassionate and fair about issues that science has not yet resolved, such as the likely potential for risk of harm associated with developments in technology. This is unrealistic; science journalists in particular tend to be extremely well informed about developments in science and technology, but cannot reasonably be expected to be more informed than the scientific community about the potential for risk, especially in situations where much uncertainty remains. This argument reinforces the relevance of both information subsidies and issues of trust to situations of uncertain risk. Both journalists and news consumers must decide which voices are most credible with respect to interpretations of uncertain evidence about likely risk, and this is especially true when the scientific evidence is mixed at the early stages of a technology.

How these dynamics will play out in the age of the Internet, in an era of rapid change in the media industry and diminished resources for investigative reporting of complex topics, is still largely unknown. The role of information subsidies, however, is almost certain to become even greater, as a much greater proportion of the information available to consumers will undoubtedly, in the future, come directly from various stakeholders via the Internet, bypassing conventional media altogether. The effective and responsible management of risk will therefore depend more heavily than ever before on the collective wisdom of citizens acting as informed information consumers. This represents a strong argument that responsible risk communication will be that which best supports such a citizenry. That does not mean that this kind of information environment will be easy to produce, however.

Media Effects Theory in the Internet Age

A distinctive stream of scholarly research literature about media, heavily empirical and more strongly rooted in social psychology than the literature considering media's more general role in society, considers the media's direct and measurable effects on public opinion and behavior. The extent to which media accounts actually influence public opinion matters to the analysis of whether and how the nature of media accounts might (in turn) matter to the management of modern democracies. Risk communication is to some extent a form of strategic communication, with different stakeholders seeking slightly different outcomes. Thus, the conclusions from research about media effects are of relevance to all of these stakeholders, from consumer advocates and proponents of strong regulation to "free market" technology advocates. Risk communicators typically represent agencies and organizations with an interest in persuading audiences that particular risks are, or are not, worthy of widespread public concern. But do the media really have that power?

The science regarding media effects may strive to be largely independent of normative considerations, but underlying ethical considerations cannot be avoided. For example, if media portrayals of violence actually promote violence in real life, what does this say about the media's ethical obligations with respect to programming incorporating violence, especially that aimed toward children whose values and behavior patterns are not fully formed? Similarly, if media stress on risk influences public perceptions of that risk, what are the media's ethical obligations with respect to an "objective" representation of risk issues? These questions beg yet other considerations of the role of the media in influencing public opinion, as well as whether it is even possible to produce an objective evaluation of risk in the face of scientific uncertainty in any given situation.

The evidence—over many decades of research and many thousands of studies relevant to the effects of media on public opinion and behavior—remains mixed. It still seems likely that media are an important influence on public opinion, not just by providing the information that serves as "fodder" to fuel public sphere debate on the issues of the day, but also having the power to shape that debate (e.g., defining some perspectives as "legitimate" and others as "illegitimate," some points of view as "mainstream" and others as "fringe," and some issues as deserving a high priority on the public agenda and others less important). However, for the most part, researchers seeking strong evidence of powerful, direct, immediate, or uniform effects from media on public opinion have been disappointed. The real picture is much more nuanced, with indirect and long-term effects that vary from individual to individual likely but often difficult to demonstrate.

Complicating any analysis of the influence of media on public reactions to risk information is the fundamental structural changes that are currently taking place in how news and information are delivered. This predates the Internet age by many decades. Fifty years or so ago, only three news networks and a secondary "public broadcasting" presence controlled American broadcast news. Most people interested in current events read their local newspaper as well. Various contemporary societies experienced somewhat different histories, but the bottom line is similar: What once were few voices has now been expanded to many more, even before the advent of the Internet. Today, researchers interested in exploring the impact of the mass media must first solve the problem of what media to study, which content among the many available sources might actually be influential, and how to isolate their influences, especially long-term effects that cannot easily be studied by designing one-shot experiments.

Of the many thousands of studies of media and its effects that have been conducted over this period, most classic approaches conceptualized a media system in which channels were of limited diversity and availability. To some extent this is still true, but there is no denying that the expansion—first of cable and satellite, and now of Internet-based news—has changed things forever. Even controlled-press countries can no longer maintain their monopolies on news, with very few exceptions. News is now more diversified, demassified, and user controlled than ever. Everything from blogs and "YouTube" opportunities mean that anyone can produce their own news, for better or for worse. The old-time editorial "gatekeepers" on which much mass communication effects research was predicated have vanished. Contemporary media effects research focuses more and more on how information consumers make choices to seek particular types of information from particular sources and channels, rather than envisioning uniform exposure to certain messages.

What we can say now about "media effects" should probably be qualified more carefully now than ever, given these developments. New approaches to researching these issues are needed. However, we have long known that the media cannot make effective propaganda "out of whole cloth," as it were. From Hitler's German propaganda team to today's Madison Avenue ad agencies and moving forward into an Internet-based information future, attempts to manipulate "mass" publics have been dependent on successful appeals to those publics' interests, values, and beliefs. Research on nanotechnology to date has confirmed that preexisting values and beliefs with respect to technology are the most likely driver of public reactions to nanotechnology, probably much more powerful than media portrayals of the technology (Kahan et al. 2009). Risk communicators can (and should) point out both nanotechnology's benefits and its risks, but powerful predilections to interpret technology as a good or as a bad influence on the direction of societal change are still with us, and still seem likely to rule the day.

How is public opinion formed in the Internet age? Mass communication effects research went through a major paradigm shift before the Internet even arrived on the scene, moving from a "magic bullet" model in which the predominant assumptions were that people would consistently be influenced by what they read and heard in predictable ways and that opinion would generally follow media representations, to a "limited effects" model in which it was recognized that public opinion is formed on the basis of many factors, including values and beliefs, not just as a simple "knee-jerk" reaction to media messages. The proliferation of channels and sources made possible by new communication technologies has made it much more difficult to study and assess what kinds of information people receive from the media, because news consumers have so many more choices available, but it has also further underscored the weakness of older theory that expected media to strongly control public opinion. Media coverage does affect, but does not determine, the public and policy agendas for society; rather, multiple institutions working together both create and modify these agendas, an idea known as "agenda building" (Lang and Lang 1983). Information consumers are active seekers and processors of messages and information, not merely passive audiences.

This is as true for news of emerging technologies such as nanotechnology as it is for information about other social, political, and scientific issues. Media messages should not be discounted but are just one factor among many in driving public opinion and policy action. Regardless of exactly how we define *media* in the Internet age, its influence is clearly more varied and diverse than it was in the past, and the active role of audience members as information seekers is more apparent. Consumers have many more information choices, as do risk communicators and risk managers advocating for particular positions. Younger generations may get most of their news from nontraditional sources, ranging from comedy to blogs and other do-it-yourself forms of media. Yet, while the present economic restructuring of the media industry may be painful for employees of media organizations in danger of losing their jobs, it is not necessarily bad for democracy if the end result is that more voices are being heard. One downside is that traditional media organizations now have fewer resources to invest in specialized journalism, such as journalism about science and technology, as well as in investigative journalism generally. If only subsidized voices are heard on these topics, consumers may be disadvantaged rather than better served.

This discussion returns us to the question of the appropriate role of the mass media and other public communication in democratic discourse. The ideal that the media will provide a marketplace of ideas from which informed citizen-consumers can choose the best ones is paralleled by the thinking behind contemporary experiments with novel forums for citizen discussion about emerging technologies; both are similarly inspired by a hope that broad public discussion has the capacity to produce better technology policy outcomes. More critical, less optimistic, views suggest that more heavily subsidized perspectives could have an unfair advantage in today's expanded marketplace, and that future risk communication on behalf of particular interests will reinforce the power of the most heavily subsidized perspectives.

On one hand, there is hope for a further democratization of power as new media begin to complement other initiatives that attempt to encourage public discussion of technology policy issues; on the other, perhaps these new alternatives will simply tend to reinforce the status quo distribution of power. If the promoters of emerging technology fully embrace the concept that a democratic consensus should underlie their efforts, and that broad public discussion is in their own interests even if it does not always "go their way," then there could be a genuine confluence of interests possible. In an ideal scenario, a combination of new and traditional media, plus the institution of other novel forums for generating public discussion (on- and off-line), may facilitate better democratic discourse surrounding a collective imagining of the future relationship of technology and society. Whether this potential will be realized, of course, remains to be seen. The emergence of nanotechnology has at least provided new opportunities to experiment along these lines.

The enhanced ability of individuals to seek out and process information from a broad variety of sources on the Internet and elsewhere is certainly a positive opportunity for the improvement of processes of democratic debate on emerging technology issues. However, ordinary citizens have many different questions, problems, and issues competing for their attention. And in a truly "demassified" media environment in which everyone seeks their own information, some have worried that the shared foundation for public discussion is weakened. Individuals seeking information on the Internet can be expected to be biased toward sources agreeing with their own perspectives, whether critical of new technology or enthusiastic about it. When most concerned citizens read the same newspapers and watched the same evening news programs, a more focused agenda for popular debate may have been created. The new media information may have created a more "fractured" society in which individuals move along very different paths and can avoid confronting opinions that challenge their own (see, e.g., Neuman 2001).

Some commentators have wondered how agenda-setting will take place in the new media environment facing us. Will audiences and publics with a special interest in technology, whether pro or con, become the primary audiences for accounts of new science and technology, while others may feel free to ignore them, focusing on other types of news instead? This has always been true to some extent; some of us read the science page while others prefer to focus on politics, business, sports, or entertainment. The further proliferation of Internet-based choices only exacerbates this tendency to focus on just those issues of most immediate personal concern. Public engagement activities focused on emerging technology tend (likewise) to be limited to people with some kind of a special interest in future technology issues (Kleinman et al. 2009). The technology we choose today will affect all of us. This remains an unsolved problem for democracy.

The nature of future audiences for science and technology news is only one concern. Will risk communicators and other public information specialists working for technology's promoters, whether corporations or universities, become the primary sources of technology-related news and information? As traditional news organizations let go increasing numbers of science specialists in response to harsh economic times, this is a distinct possibility. The consequences for public engagement in choices about technology are not clear. So long as public acceptance of nanotechnology remains high, the absence of neutral forums for information and discussion is not a major issue. However, this may not always remain the case. For nanotechnology, many observers have speculated that as the technology becomes more prominent in life science applications such as food packaging, medical care, and agriculture, the current level of acceptance may fade. Increasing production of nanoscale materials and products incorporating nanoparticles could expose more and more workers to risks, with important consequences for public acceptance. Present attempts at creating new forums for public engagement represent attempts to be proactive in anticipation of potential controversy, but may not last without ongoing sources of funding.

Museums and Science Centers

Fully exploring the role of other institutions in disseminating information about science and technology is outside the scope of this book. The present discussion would not be complete, however, without at least briefly recognizing the role of other social institutions in disseminating information about science and technology to out-of-school adults. Once someone has stopped being a student, how does that person remain connected to ongoing developments in technology and its potential impact on society? The ordinary news media may not accomplish this, except for those individuals motivated to seek out and process the information via new media or other sources. Journalists, whether working in old or in new media, do not always think of their job as empowering or even educating citizens, but (more modestly) as simply informing them. As a result, some envision an expanded role for science museums and science centers in this regard (Bell 2008; Priest 2009).

Historically, many science museums have avoided engaging in policy discussions, seeing their role as encouraging interest in the underlying science instead (much as journalists limit their own role to informing). Yet scholars have increasingly argued that science and science policy cannot always be neatly separated. Further, science museums and science centers have the respect of those who frequent them and seem in a good position to both behave, and to be perceived, as neutral arbitrators of science and technology information as opposed to representing particular interests. In the United States, for example, the Nanoscale Informal Science Education, or NISE, Network sponsors a national event known as NanoDays that is designed to appeal to both young people and adults at over 200 locations in science museums, science centers, universities, and other research organizations, seeking to engage the public and raise awareness of nanotechnology's role in our future.

Not only do these and related special events have many participants, but science museums and centers, many of which have included exhibits on nanotechnology and other emerging technologies, exist in many cities. And some of these institutions are taking the lead in developing programs that address ethics and policy and that target adults as well as schoolchildren. For example, in November 2009, the Science Museum of Minnesota sponsored a forum on privacy, civil liberties, and nanotechnology, exploring the idea that nanoscale technology may enable new ways to track threats but at the same time also present new challenges to maintaining the balance between an individual's right to privacy and community safety. The Museum of Science, Boston, has sponsored a broad range of events on specific developments in nanotechnology, from quantum computers to cancer screening, and at least one forum on whether nanotechnology consumer products are safe.

These are only a few examples of museum efforts in the area of nanotechnology and society. Universities with significant nanotechnology and society programs, including the University of Arizona and the University of California at Santa Barbara, are also experimenting with new outreach vehicles. Like other special public engagement events, these events, programs, and exhibits will not reach everyone. Expanding the *attentive public* (Miller and Pardo 2000) for science and technology will remain an ongoing challenge. However, this is a challenge that needs to be addressed. The technology we adopt today, in applications ranging from alternative energy to medicine, from electronic communication to food production and packaging, and from environmental remediation to transportation—will shape our social world tomorrow. And much of our future new technology in these and many other areas is likely to incorporate nanotechnology elements, if present predictions are anywhere near correct. Wise management of these opportunities will undoubtedly require ongoing efforts to keep the public aware of both risks and benefits.

PERSPECTIVE: WHAT HAVE THE MASS MEDIA BEEN REPORTING ON NANOTECHNOLOGY RISKS?

Sharon M. Friedman
Brenda P. Egolf

The mass media often play an important role in signaling to the public that both benefits and risks result from emerging technologies. In the case of nanotechnology, studies have shown that most media coverage has focused on the positive contributions that nanotechnology is expected to make to the economy, medicine, new materials and information technologies, among others (Lewenstein et al. 2005; Stephens 2005; Nisbet and Scheufele 2009). In contrast, there has been far less discussion in the media about nanotechnology risks.

After reviewing 10 years of coverage in 29 major newspapers and two wire services in the United States and the United Kingdom, we found that less than 6% of about 6200 general articles on nanotechnology included any discussion of health or environmental risks. Yet, potential threats of long-term illnesses or environmental contamination from nanotechnology products or manufacturing are prime examples of the types of risks that could be intensively covered by the media and then have negative economic or reputational effects on companies, industries, and even other emerging technologies.

THE STUDY

We wanted to study newspaper coverage of nanotechnology risks to see what laypeople who knew little or nothing about the field and were not searching for information about it could read if they were interested. We used two wire services as a substitute for some broadcast coverage, because television and radio newscasts often take information from the wire services and use it on air. In addition, we wanted to view this risk coverage over a long period to look for long-term trends rather than just the effects of one or two incidents. To do this, we reviewed more than 2200 articles between 2000 and 2009 in 20 U.S. and 9 U.K.[1] newspapers and the Associated Press and United Press International wire services that were in the Lexis-Nexis

[1] A newspaper from the Republic of Ireland, *The Irish Times*, was included because of nanotechnology research occurring in that country. For convenience, results from this newspaper are included under the U.K. label, although the Republic of Ireland is not part of the United Kingdom.

Academic database. We used standard search terms: *nanotechnology*, and *risk or problem* or *issue* or *concern* or *toxicity* or *safety*, and *environment* or *health*. When we found relevant articles, we analyzed their content using a coding document that included, among many items, the number of articles, types of risk, generators of risk news, regulatory coverage, and information sources used by journalists.

To try to compare the number of risk articles to those on nanotechnology benefits and other issues, we conducted a very general search in Lexis-Nexis Academic using only the term *nanotechnology* in the same newspapers and time period. We found more than 6200 articles on the general topic but did not review these articles for relevancy, so some articles might not be directly applicable to nanotechnology issues.

NANOTECHNOLOGY RISK ARTICLES BY COUNTRY AND YEAR

Even with this extensive search, we only found 367 articles that mentioned nanotechnology health or environmental risks—248 from the United States and 119 from the United Kingdom. As mentioned above, this is about 6% of the more than 6200 general nanotechnology articles and about 17% of the 2200 risk articles retrieved. The trend over time in number of articles varied between the two countries. The number of U.S. articles increased between 2001 and 2006, rising in uneven spurts from 4 to 57. At first it appeared as if 2006 with its large number of articles indicated a new level of U.S. interest in nanotechnology risk coverage, but this turned out not to be the case because coverage dropped after that. Only 34 articles appeared in both 2007 and 2008, and this number dropped to 17 in 2009 (see Figure 7.1). In the United Kingdom, 20 or more articles appeared in 2003, 2004, and 2008; the rest of the years saw articles in the teens or lower. The largest number of U.K. risk articles was 25 in 2004. Most of the articles were written by general reporters, but about 38% in the United States and 35% in the United Kingdom were written by journalists who specialized in science, environmental, health, or technology issues.

EVENT REPORTING

While the small number of articles was of concern, another somewhat disturbing factor was that most of the coverage in both countries was about news events. About 69% of the U.S. and 58% of the U.K. articles were news stories based on the release of a study or report, an event,

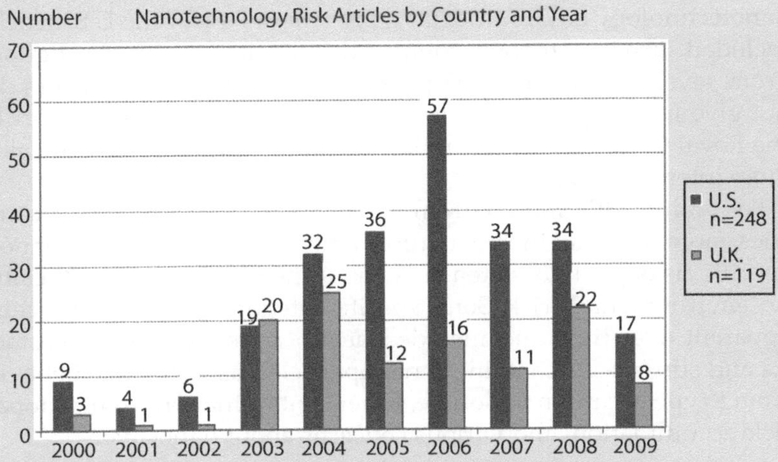

FIGURE 7.1
Nanotechnology risk articles by country and year.

or statements by well-known people. In the U.K. coverage, for example, a commentary from Prince Charles and a major report from the Royal Society and the Royal Academy of Engineering resulted in the largest number of articles, which appeared in 2004. In 2008, news coverage of a scientific study and a government commission's report brought about the second largest number of U.K. articles. News reporting is expected in covering emerging technologies; inclusion of feature, explanatory, and investigative articles is needed because they offer more information for better reader understanding of and knowledge about scientific and technological issues.

TYPES OF RISK COVERAGE

Coverage of health risks, found in 74% of the articles, exceeded that of environmental risk coverage in 67%. More than half of the health risk articles in both countries appeared between 2006 and 2009, perhaps indicating more scientific or government activity or media interest during that period. Somewhat similar percentages of articles discussed health risks in both countries, 75% and 71% for the United States and United Kingdom, respectively.

We found a larger difference in the two countries' coverage of nanotechnology environmental risks. About 70% of the U.S. articles included such information, compared to 61% in the United Kingdom. From several measures, it appeared that the U.K. newspapers did not give much space to environmental risks, although a report by the Royal Commission on Environmental Pollution in 2008 focused more interest on this area.

For both health and environmental risk articles in the two countries, more than half included information only about general or nonspecific risks, such as statements about the possibility of a "health" or "environmental" risk. Some specific categories of concern became apparent over time and included health risks from using cosmetics and sunscreens containing nanoparticles, asbestos-like behavior from a type of carbon nanotube, water contamination from nanoparticle accumulation, and various concerns about nanosilver.

TOP GENERATORS OF RISK NEWS

A risk event can contribute to public concern over an emerging technology such as nanotechnology, particularly if there is an abundance of media coverage during the same time period, indicating the importance of the event. We identified 10 events that generated a cluster of newspaper articles about various nanotechnology risk issues (listed chronologically):

- April 2003: Prince Charles wrote a letter that raised a vision of a "grey goo" disaster scenario of nanotechnology and asked the Royal Society to discuss the new technology's "enormous environmental and social risks," plus responses to the article (six articles).
- March 2004: U.S. scientist Eva Oberdorster announced her study showing brain damage in fish from buckyballs (six articles).
- July 2004: Prince Charles wrote an article in *The Independent on Sunday* again sounding an alarm about the risks of nanotechnology development, plus responses (eight articles); the Royal Society and Royal Academy of Engineering released a major report calling for tighter regulations and better safety assessments to minimize health and environmental risks from nanoparticles (seven articles).
- January 2006: The Project on Emerging Nanotechnologies (PEN) of the Woodrow Wilson International Center for

Scholars released a report by J. Clarence Davies that called for more aggressive regulatory oversight and new resources to manage the potential adverse effects of nanotechnology (five articles).

- April/May 2006: Bathroom cleaning product, "Magic Nano," was recalled in Germany after users suffered health problems, plus its resolution (six articles).
- May/June 2006: Environmental groups petitioned the U.S. government for stronger regulation of skin creams and sunscreens containing nanoparticles, plus other articles on nanomaterials in skin creams and sunscreens (seven articles).
- November 2006: U.S. Environmental Protection Agency decided to require manufacturers using nanosilver to provide scientific evidence of no harm to waterways or public health.
- December 2006/January 2007: Berkeley, CA, City Council agreed to regulate businesses that make or use nanoparticles, plus followup (six articles).
- May 2008: U.K.–U.S. scientific study found long carbon nanotubes caused health risks similar to asbestos (seven articles).
- November 2008: The Royal Commission on the Environment issued a report calling for more safety testing and tighter regulations of nanomaterials (seven articles).

This list includes topic groupings in either or both countries that occurred at a particular time. The number of articles for each topic shows that none reached a level of coverage significant enough to generate concern in lay readers. In addition, we found that this type of aggregated newspaper and wire service coverage was the exception rather than the rule. Particularly in the United States most topics or events were reported by only one newspaper or wire service at a time, so there were many topics, such as a scientific study, that appeared in only one article. Other topics were scattered, appearing in one or two articles in one month and then reappearing several months later in another article or two.

REGULATORY ISSUES

We found that discussions about regulation were not as frequent as those about health and environmental risks but still played a prominent role in discussing nanotechnology risks. Regulations' issues were discussed in 44% of the 367 articles and became most prominent in 2006, when they were in 65% of the U.S. articles. In succeeding years, they were mentioned in 50%, 41%, and 47% of the U.S. articles. In the United Kingdom, there was an earlier discussion of regulation, and 3 years in which regulations' issues were discussed in more than 50% of the articles: 2004, 2005, and 2008. Most of the early articles, as indicated above, dealt with comments by Prince Charles and the major report by the Royal Society and the Royal Academy of Engineering. In 2008, the report by the Royal Commission on Environmental Pollution again raised the regulation issue.

Three-quarters of the regulation articles focused on calls for new or tightened regulations with the two leading reasons for these calls in both countries being to protect public health and safety and to protect the environment. The third leading reason was that existing rules and regulations did not apply to nanotechnology.

In the United States, a small group of environmental organizations, which included the ETC Group and the International Center for Technology Assessment, called most frequently for new and tightened regulations, followed by the Wilson Center's PEN. In the United Kingdom, scientific and engineering societies as well as individual scientists were most often identified as those seeking more or better regulation.

INFORMATION SOURCES

Public opinion leaders can play a crucial role in either hyping or easing public concerns about risks. The credibility of these leaders and the levels of trust accorded to them play a major role in the impact their utterances have about a particular risk issue, particularly as they are reported by the mass media.

We grouped information sources used in articles in our study by whether they raised concerns about nanotechnology risks or whether they responded to risk issues by discussing information that would reassure readers that nanotechnology risks were not serious or could be managed. Some sources were counted in both the raised and responded categories when they did that in the same article, which happened fairly often.

There were 623 sources who raised risks in our 367 articles, an average of about 1.7 per article. The leading risk raisers were not environmental organizations, as some members of the nanotechnology community had expected, but instead were university scientists and engineers. They were followed by nanotechnology organizations, primarily the Wilson Center's PEN. A lower number of sources,

436, responded to the risk issues, for an average of 1.2 per article. University scientists and engineers also dominated this source group, followed by industry and government organizations.

Most of the risk raisers used by the media as sources probably would be considered trustworthy by members of the public in their respective countries. This trust would occur because of the sources' roles as university scientists and engineers, rather than because of their individual identities. The reporters used these individuals frequently, probably because of their research prominence and stature in the field, their membership on major advisory committees, and their willingness to talk about a highly technical subject in an understandable way.

CONCLUSIONS ABOUT OVERALL MEDIA COVERAGE

The number of articles we found in this study was surprisingly low for a 10-year period, averaging only about 37 articles per year in 29 publications and two wire services. This appears particularly low when compared to other studies of media coverage of risk subjects over time where averages ranged from 56 to 91 articles per year (Desilva et al. 2004; Holliman 2004; Trumbo 1995). Because this study's goal was to evaluate articles that any layperson had the opportunity to read in newspapers, it did not include other stories that might have appeared in the broadcast media, in general or scientific magazines, or on the World Wide Web. It also did not include articles in newspapers that were not part of the study sample, so this low number might be augmented by other sources. In particular, numerous Web sites have sprung up to cover nanotechnology news. Many surveys have shown, however, that there is a lack of public awareness of nanotechnology (Cobb and Macoubrie 2004; Peter D. Hart 2006, 2008; Siegrist et al. 2007; Satterfield et al. 2009), which indicates that few laypeople are actively seeking nanotechnology information, even on the Web.

Part of the lack of risk coverage found can be blamed on downsizing in the print media. Some U.S. reporters who were interviewed for this study indicated that nanotechnology is too specialized a topic to attract the public, and it is not local enough to warrant limited space. In addition, they noted, nothing new was happening concerning nanotechnology risk issues.

Even in the small number of articles that we did find, there were problems. The event reporting that dominated the nanotechnology risk coverage did not provide continuity to help readers understand the evolving nature of nanotechnology risk research or regulation. In addition, because the articles often lacked in-depth coverage, they did not put risk information into context. A lack of contextual information can confuse readers, particularly concerning technical subjects.

Another issue of concern is the loss of specialized newspaper reporters who can write about this complicated subject. In the United States during the last few years, a large number of science and environmental writers have taken buyouts or been laid off from their newspapers, including almost all of those who wrote the U.S. nanotechnology articles in this study. This loss has already had an adverse impact on U.S. nanotechnology risk coverage, particularly in the influential *New York Times* and *Washington Post*.

For these reasons, the chances for explaining a future harmful nanotechnology event through articles in newspapers and wire services are not bright. If a threat to public health and safety or the environment from a nanotechnology product were large enough to warrant extensive mass media attention, it probably would not be explained well enough to help readers understand the situation. Furthermore, given the shrinking amount of space and growing local news orientation of many U.S. newspapers, it is unlikely that a national or international story about a nanotechnology risk would be lengthy or have much depth.

In the future, there probably will be diminishing opportunities for U.S. and U.K. citizens to find articles about nanotechnology health and environmental risks in their newspapers. So where will readers find them? More and more nanotechnology news, mostly about benefits but occasionally about risks, is migrating to the Web, where members of the nanotechnology community, interested elites, and scientifically attentive members of the public can find it. However, people have to *seek out* this information on the Web; it will not just appear on a newspaper page that someone peruses while eating breakfast. Only very interested members of the public will probably do such seeking, leaving the majority of citizens naive about potential nanotechnology risks. This is exactly the kind of situation that could generate fear, anger, and distrust among members of an uninformed public should a serious nanotechnology accident or long-term health or environmental effect occur.

ACKNOWLEDGMENTS

This research was funded by a grant from the National Science Foundation to the Harvard-UCLA Nanotechnology and Society Network, NCES CNS SES #0531146, and by a Materials Research Science and Engineering Center grant from the Pennsylvania Department of Community and Economic Development, contract #C000007361. The opinions are those of the authors and not the granting agencies.

We would like to thank the students who worked on this project with us and Lehigh University's Center for Advanced Materials and Nanotechnology for its support.

REFERENCES

Cobb, M.D., and J. Macoubrie. 2004. Public perceptions about nanotechnology: risks, benefits and trust. *Journal of Nanoparticle Research* 6: 395–403.

Desilva, M., M.A.T. Maskavitch, and J.P. Roche. 2004. Print media coverage of antibiotic resistance. *Science Communication* 26: 31–43.

Holliman, R. 2004. Media coverage of cloning: a study of media content, production and reception. *Public Understanding of Science* 13: 107–130.

Lewenstein, B., J. Gorss, and J. Radin. 2005. The salience of small: nanotechnology coverage in the American press, 1986–2004. Paper presented at the annual convention of the International Communication Association, New York, May.

Nisbet, M.C., and D. Scheufele. 2009. What's next for science communication? Promising directions and lingering distractions. *American Journal of Botany* 6: 1–12.

P.D. Hart Research Associates, Inc. 2006. Report findings: attitudes toward nanotechnology and federal regulatory agencies. *Project on Emerging Nanotechnologies Report*, September 19. Retrieved June 2010 from www.nanotechproject.org/file_download/files/HartReport.pdf.

P.D. Hart Research Associates, Inc. 2008. Awareness of and attitudes toward nanotechnology and synthetic biology. *Project on Emerging Nanotechnologies Report*, September 16. Retrieved June 2010 from www.nanotechproject.org/process/assets/files/7040/final-synbioreport.pdf.

Satterfield, T., M. Kandlikar, C. Beaudrie, J. Conti, and B. Harthorn. 2009. Anticipating the perceived risk of nanotechnologies. *Nature Nanotechnology* 4: 752–758.

Siegrist, M., C. Keller, H. Kastenholz, S. Frey, and A. Wiek. 2007. Laypeople's and experts' perception of nanotechnology hazards. *Risk Analysis* 279(1): 59–69.

Stephens, L. 2005. News narratives about nano S&T in major U.S. and non-U.S. newspapers. *Science Communication* 27: 175–199.

Trumbo, C. 1995. Longitudinal modeling of public issues: an application of the agenda-setting process to the issue of global warming. *Journalism and Mass Communication Monographs* 152: 1–57.

References

Bell, L. 2008. Engaging the public in technology policy: a new role for science museums. *Science Communication* 29(3): 386.

Boykoff, M.T., and J.M. Boykoff. 2007. Climate change and journalistic norms: a case-study of U.S. mass-media coverage. *Geoforum* 38(6): 1190–1204.

Friedman, S.M., S. Dunwoody, and C.L. Rogers. 1986. *Scientists and Journalists: Reporting Science as News*. Glencoe, IL: Free Press.

Gandy, O.H. 1982. *Beyond Agenda Setting. Information Subsidies and Public Policy.* Norwood, NJ: Ablex.

Kahan, D.M., D. Braman, P. Slovic, J. Gastil, and G. Cohen. 2009. Cultural cognition of the risks and benefits of nanotechnology. *Nature Nanotechnology* 4(2): 87–90.

Kasperson, J.X., R.E. Kasperson, N. Pidgeon, and P. Slovic. 2003. The social amplification of risk: assessing fifteen years of research and theory. In *The Social Amplification of Risk*. Eds. N. Pidgeon, R.E. Kasperson, and P. Slovic. Cambridge: Cambridge University Press, 13–46.

Kleinman, D.L., J.A. Delborne, and A.A. Anderson. 2009. Engaging citizens: the high cost of citizen participation in high technology. *Public Understanding of Science*. doi: 10.1177/0963662509347137.

Lang, G.E., and K. Lang. 1983. *The Battle for Public Opinion: The President, the Press, and the Polls during Watergate.* New York: Columbia University Press.

Lasswell, H.D. 1948. The structure and function of communication in society. In *The Communication of Ideas.* Ed. L. Bryson. New York: Harper and Brothers, 37–51.

Marx, K. 1999. *Capital: An Abridged Edition*. Trans. D. McLellan. New York: Oxford University Press.

Miller, J.D., and R. Pardo. 2000. Civic scientific literacy and attitude to science and technology: a comparative analysis of the European Union, the United States, Japan, and Canada. In *Between Understanding and Trust: The Public, Science and Technology.* Eds. M. Dierkes and C. von Grote. New York: Routledge, 131–156.

Neuman, W.R. 2001. The impact of the new media. In *Mediated Politics: Communication in the Future of Democracy.* Eds. W.L. Bennett and R.M. Entman. Cambridge: Cambridge University Press, 299–320.

Powell, M., and M. Colin. 2008. Meaningful citizen engagement in science and technology: what would it really take? *Science Communication* 20(1): 126–136.

Priest, S. 2009. Science and technology policy and ethics: what role should science museums play? *Museums and Social Issues* 4(1): 55–65.

Wright, C.R. 1986. *Mass Communication: A Sociological Perspective,* 3rd ed. New York: Random House.

8

Lessons and Future Challenges

In many ways, nanotechnology in the public sphere has not developed as anticipated, in terms of public discussion, public perception, or public reception. Little or no evidence is available to suggest that science fiction scenarios of impending disaster associated with nanotechnology have been generally embraced, or that concerns about one form of nanotechnology are thoughtlessly applied to other forms in terms of popular perception. Most prominent among the unrealized prophesies, and quite contrary to some expectations, is the general failure of nanotechnology to generate visible public concern. Nanotechnology has been associated with relatively little public controversy, even though some forms are now believed to involve very tangible hazards to health or to the environment. Debates involving whether and how to regulate nanomaterials have sometimes been quite heated, but these have often been primarily confined to discussions among regulators, advocates, and other stakeholders, not prominently spilling over into the public sphere. Risk coverage in the news media even seems to be on the wane. Little in the way of visible or virulent argument or protest from nonexpert publics seems to have emerged, compared to the reactions and events that seemed to characterize previous developments in areas such as nuclear power, agricultural biotechnology, or stem cell research.

Although it is tempting to suggest that the greater attention paid to the communication and social scientific aspects of nanotechnology in comparison to previous waves of technological innovation might be the reason for its public reception having been relatively calm, we have no direct evidence of this. Many observers have speculated that new developments in agriculture and food processing that combine nanotechnology and biotechnology (to produce *nanobiotechnology*) will prove more controversial, but even this has yet to occur. Others point to the prospect of human enhancement via nanotechnology as the likely site of future controversy, although this remains a very futuristic vision.

Yet others have suggested that contemporary new developments in synthetic biology, which (because they also involve the manipulation of matter at the molecular scale) can be considered a close cousin of nanotechnology, will be much more controversial. Synthetic biology techniques create desired segments of DNA "from scratch," based on artificially generating known sequences of DNA base pairs. *Synbio* represents the next great wave of human control over natural processes and is expected to have important implications in areas ranging from the production of everything from pharmaceuticals to biofuels. Yet some fear these techniques. While offering promise in many areas, they can equally be used to create undesired materials (e.g., by making diseases more virulent, or even by recreating past threats now under control, such as smallpox). Terrorist abuses have been all too clearly envisioned. And yet, as with nanotechnology, seemingly little public attention has followed these developments, and little public concern seems to have attached to them.

The concept of social amplification of risk suggests that some risks become socially amplified, or exaggerated, as relevant information travels through society, sometimes propelled by attention from prominent institutions such as the news media, while other risks are attenuated, receiving little attention (setting aside the thorny issue of which risks might actually deserve society's attention, or not). Yet this idea has little to offer in terms of predicting which risks will become amplified and which, on the contrary, will become attenuated. It may be that amplification "waves" require a "perfect storm" scenario in which actions and reactions by multiple institutional stakeholders converge together to yield a spike in public concern and resistance. Alternatively, as seems equally likely, hazards involving culturally sensitive areas, such as the food we consume in our bodies or our individual and familial genetic heritage, carry more cultural resonance and thus are more susceptible to amplification effects.

Most likely, both processes—that is, social amplification effects requiring particular confluences of circumstances, alongside cultural resonance effects that are attached to technologies emerging in especially sensitive areas—occur simultaneously, and both sets of dynamics are important to understanding popular reactions. Conversely, technologies that promise to address important cultural priorities, such as health care, may be received as less controversial, a phenomenon that might be viewed as positive (rather than negative) cultural resonance. This clearly applies in some cases of nanotechnology. In both types of cases, whether amplification or resonance matters the most, institutional activities (including the representations produced by news media) still matter as well, but some technologies appear more susceptible to amplification than others, and some clearly strike either positive or negative chords with respect to broadly shared cultural values.

The diminished capacity of the present-day news system—strapped for resources and bleeding both jobs and advertising dollars—to provide meaningful information adequate to support informed public debate is a source of special concern here. If an important set of risks should arise that do not strike a compelling cultural chord, today's societies are largely dependent on the media to point this out. Even though amplification of some risks due to perfect storm convergence of varied stakeholder and other institutional interests is a possibility, attenuation of other risks, particularly those that do not have great cultural resonance, is also a possibility. This is doubly true absent vigorous science journalism. In the case of nanotechnology, some social observers may have anticipated social amplification, but the current reality appears to be attenuation—complacence rather than concern. Naturally, future events may, alternatively, either demonstrate that concern was not actually warranted, or that risks should have gotten more attention sooner. Substantial uncertainty remains; only hindsight can sort out how much attention this novel set of hazards actually might deserve, or how best to weight its risks and benefits. The appropriate role of risk communication is, in part, to empower and guide us to make wise decisions under uncertainty.

And, of course, nanotechnology also promises myriad future benefits that matter to people: enhanced health-care options, more efficient fuel use, new materials for use in everything from alternative energy generation (wind turbines, solar cells) to sports (tennis rackets, skis) to cosmetics. Here the analogy with biotechnology is especially instructive, however. Biotechnology promised, and continues to promise, new solutions to the problems of world hunger and rural poverty. The technology available may be up to the challenge, but more than goodwill and sound science are necessary to realize that potential. Present applications of agricultural biotechnology, for example, are heavily concentrated in the production of only a few crops grown on very large scales (soybeans, corn, and cotton). Its application to solve local problems of nutrient deficiency (e.g., "yellow rice" as a solution to vitamin A precursor deficiencies that lead to blindness in some parts of Asia) or the need for crops that can be grown on otherwise marginal agricultural land (e.g., in parts of Africa) remain underdeveloped. Both biotechnology and nanotechnology have the potential to contribute to the solution of nutritional, health, and economic problems around the globe. Biotechnology's potential for this clearly remains unrealized. Will nanotechnology follow the same path?

Does the attenuated popular perception of nanotechnology-related risk mean that nanotechnology risk communicators have little work to do? At present, they may have little *crisis* communication work to do, something that could certainly change in a heartbeat. Yet there is still much *risk* communication work to be done. Bearing in mind that risk communicators represent a variety of different interests, not just those of a technology promoter, risk communicators have a particular role in encouraging recognition and appropriate regulation of nanotechnology's risks, again, in short, enabling wise decision making on the part of society. Progressive commercial and industrial interests may proactively support reasonable regulation, which may give them some level of legal protection with respect to worker and consumer protection. Here, risk communicators representing these interests could well find themselves on the same side of the fence as those that represent worker and consumer interests, advocating for legislative and regulatory action.

One relevant and important lesson from the history of the previous societal debates over technology, from biotechnology to nuclear technology, is the central role of trust in resolving technical disputes. Everyone from ordinary nonexperts (that is, those who are sometimes referred to as "lay" citizens) to accomplished technical specialists is dependent on advice from people with other perspectives. The nuclear engineer is not an expert in environmental impact, nor is the environmental advocate an expert in the management of a nuclear facility. Each player must make decisions on the basis of trust (or distrust) in other players; while sometimes lumped under heuristic or cue-based decision making, implying more superficial analysis, this dynamic is actually a fundamental aspect of human social behavior that can be highly rational. Understanding the risks and hazards posed by nanotechnology, and communicating these to a broad range of expert and nonexpert publics effectively, will require not only knowledge of emerging research and expert opinion but also awareness of the particular role of social trust in such contexts. Building trust across disciplines and institutions and working to keep channels of communication open are wise investments in the future, especially if nanotechnology-related risks and hazards should erupt as future controversies, but even if they do not.

Media have considerable power to contribute to informed public debate, even to set the agenda for that debate, but they do not fully determine the direction that debate will take, especially in today's new media environment. In today's world of expanded media capacity and truly global communication patterns, but a diminished role for science journalism as economic pressures shrink newsroom power, this is even truer. The traditional mass media remain crucial channels for the dissemination of news and information about contemporary developments, including those in nanotechnology, its benefits, and its risks, but fewer journalists are available who can interpret this information appropriately. Most citizens today have access to a near-infinite array of sources via the Internet, but many of these sources represent stakeholders rather than neutral resources. The future of journalism in general and science journalism in particular is unclear: On the one hand, the future may well be an era in which fewer specialized journalists who can communicate technical information to a broad range of publics exist; on the other hand, for those who are motivated to seek it out, the Internet will continue to provide a wealth of technical information, but information primarily accessed by interested publics.

At one time in recent history, the "information gap" most feared had directly to do with access to Internet-based information. This is still an issue but may be less important today as Internet access becomes closer to universal. The new gap looms as one separating the interested and aware consumer of technical news who actively seeks information from the casual and relatively passive observer who does not. We cannot all be experts in everything. Just as our decision making must inevitably depend on trust-based choices, our information consumption cannot be infinite. Information overload, long recognized as a hazard of the information society, is easily achieved in today's information-rich environment. The emergence of nanotechnology has spawned new beginnings as we experiment with new alternatives for public discussion and deliberation, such as the Citizens' Forums experiments described in this volume. The role such discussions might play in future policy making remains imperfectly defined. And as traditional or legacy media outlets cut back on their coverage of science, the role of risk communication specialists will become proportionately more influential.

The risk perception and risk communication research communities and associated communities of practice have the opportunity to contribute to wise public decision making about the management of the risks of nanotechnology, as for all technologies, by pointing the way toward issues requiring attention. This will require proactively seeking to understand why people understand risks in the way that they do; understanding the role of trust in determining opinions about risk—a dynamic in play for experts just as certainly as for members of other publics; and understanding what people most want and most fear from technology—that is, the role of social values and cultural beliefs in determining why some technologies are embraced while others meet with resistance. This is not to say that understanding of these dynamics can neutralize differences (including the unequal distribution of power in society) or erase risks (including risks that are inequitably distributed, such as risks that affect workers but not management, or risks that differentially affect different kinds of neighborhoods, often raising issues of environmental justice). Even so, sensitivity to issues of trust and values as these influence public perception are central to effective risk communication, a two-way process.

PERSPECTIVE: THE ETHICS OF RISK COMMUNICATION FOR NANOTECHNOLOGY

Paul B. Thompson

The word *ethics* can be both singular and plural. As a plural noun, ethics are norms, standards, and expectations that circumscribe, direct, and possibly motivate conduct appropriate for a given situation. Among professional groups, ethics are rules and codes essential to the performance of skills, tasks, or proficiencies thought to be peculiarly characteristic of the profession. As a singular noun, ethics is a domain of discussion, discourse, and theory devoted to the critical analysis of both individual and social conduct. In academic circles, it is sometimes characterized as a subfield of philosophy, though academic programs in ethics increasingly tend to be interdisciplinary. The two meanings are not unrelated, however. Philosophers are interested in what nonphilosophers take ethical conduct to require, and real-life situations can plunge people into quandaries where thinking critically is the only way to behave ethically.

Stated succinctly, the ethics of risk communication for nanotechnology involves navigating the Scylla of scaring the pants off people and the Charybdis of shilling for industry. A less succinct account would be more helpful in the present context, and what follows offers a few additional points to help triangulate the course. There are at least two broad themes to unpack. First there are the questions of what would count as scaring people needlessly, on the one hand, or pandering for the developers of nanotechnology, on the other. Second, there are matters that have complicated the task of risk communication markedly. Even knowing where one should go is no guarantee that one will get there. The unpacking that follows in this chapter is best read as occupying the intersection space between the professional ethics (p.) of risk communication and the ethics (s.) of emerging technology.

One does not need a philosopher to explain when and why either course described above would be unethical. Apprising people of dangers that they should be aware of is a proper function of risk communication, and there is nothing inherently unethical in doing public relations work for high-technology firms developing nanotechnology. One strays into ethically troublesome territory when the communication activities one undertakes in either capacity become misleading. One is positively immoral when the miscommunication becomes tantamount to lying. Philosophers can wax eloquently on mendacity. For example, is it necessary that one's *intention* is to mislead, or can one stop short of moral turpitude by insuring that even as one messages in a manner that is sure to misdirect one's audience, one's heart remains pure? Questions about what a nanotechnology risk communication message should try to achieve may be more relevant than such musings on the ethics of truth-telling in the present context.

A venerable tradition of risk communication in public health contexts stresses attitude formation and behavioral change. Over the last 50 years, major efforts have been undertaken to apprise nonscientists of the dangers associated with smoking, with unprotected sex, with failing to buckle one's seatbelt, with sunbathing, and with binge drinking. The list can go on and on. Sometimes these campaigns are spectacularly successful. A report from 1977 describes an effort to educate women of the risks associated with leaving an intrauterine device (IUD) in place after conception. The authors indicate that fatalities from this condition fell to zero over a 5-year period (Cates et al. 1977). Other efforts that have been less successful have precipitated a line of scholarship into the way that risk communication messages can fail because they threaten their recipient's self-esteem, or initiate avoidance rather than appropriate behavioral change (Lapinski and Boster 2001; Cho and Witte 2005).

Although one might explore some subtleties in this tradition that will be ignored in the present account, we may cut to the chase by noting that this tradition of risk communication studies operates under the assumption that scientists have it right, and the job of risk communication specialists is to effect a media campaign that persuades people to change their behavior accordingly. Scaring the pants off of people would, in this view, be understood to mean scaring them in a way that is either not material to bringing about the change, or perhaps even scaring them so much that for complex psychological reasons they are incapable of making the desired change. People who are trained in this school of risk communication are unlikely to question whether the scientists might be wrong, or whether a risk communication message designed to implement the behavioral change indicated by the science could be unethical.

They do, however, have some idea of what shilling for industry might involve. David Michaels' book *Doubt Is Their Product* details a number of case studies from workplace safety standards to secondhand smoke in which industry spokesmen induced skepticism about scientific studies of risk in order to forestall regulatory policies that would have increased their cost of doing business (Michaels 2008). Carl Cranor's book *Toxic Torts* shows how industry advocates can exploit uncertainty in the courtroom to avoid liability judgments, even in cases where the scientific consensus on the toxicity of a substance is strong (Cranor 2006). In these cases, shilling for industry means promoting the interests of for-profit companies against a well-established scientific consensus. The consensus holds that risks exist and that action to curtail or limit them is warranted. Scientists are the good guys in all these stories, and ethical risk communication means telling the story that scientists want told.

There is a somewhat different line of descent for risk communication efforts that focus on assuring the public that technologies are safe. One of the topics here would certainly be nuclear power. Prospects for using heat from the decay products of nuclear fission to boil water for the generation of electrical power were noticed in the 1930s, and by the 1970s a number of general approaches had been developed. However, even as plants began to be sited in the United States during the 1960s, concern over the risks of nuclear power began to mount. By 1979, when a widely publicized accident led to the closure of the Three Mile Island plant near Harrisburg, Pennsylvania, a number of popular books opposing nuclear power had been published (Cook 1982). Nevertheless, engineers associated with the development of nuclear power took the view that opposition to nuclear power plants could not be justified on rational grounds (Starr and Whipple 1980; Lewis 1990). Opposition to nuclear power began to be blamed on a failure to undertake proper risk communication (Slovic 1986, 1987).

Opposition to nuclear power was not based solely on safety concerns, however. Although popular portrayals such as the film *The China Syndrome* and the accident at Three Mile Island had undeniably important effects on the eventual fate of nuclear power, many opponents stressed the need for less centralized systems for generating and distributing electrical power, and energy sources that were not coupled to the defense industry (Lovins 1977). Thus, concerns about the risks of nuclear power were intermingled with a broader social movement in opposition to it (Barkan 1979). This pattern of safety concerns intermingled with non-safety-oriented political goals was repeated in the surge of opposition to genetically engineered seeds (Thompson and Hannah 2008). As had been the case for nuclear power, public concern about genetic engineering in agriculture (genetically modified organisms, GMOs) was widely seen as irrational by its advocates (Miller and Conko 2001; McHughen 2008). And again, a failure of risk communication through the media was hailed as the culprit, and better risk communication was recommended as the appropriate response, often with little appreciation of what risk communication could reasonably accomplish (Priest 2000).

In comparison to risk communication efforts launched in support of public health efforts, journalists and communication professionals who found themselves in the midst of debates over nuclear power or GMOs had a far less certain status. For one thing, it was not entirely clear that the scientists were the good guys. Advocates of these technologies may well have been substantially correct in their understanding of relative safety concerns, especially as these technologies were compared to alternatives such as coal-fired electrical plants or more toxic chemical pesticides. However, while a communication effort focused narrowly on the scientific analysis of health hazards from nuclear power or GMOs might not have been stating outright falsehoods, it would hardly have been responsive to the broader substance of the debate. In responding to broad-based movements of social resistance by narrowing the scope of discourse to issues of technical risk, one might well have been accused of changing the subject. For their part, opponents of biotechnology learned that there was little point in staying strictly on point, either. They could have more success by exaggerating the risks and putting proponents on the defensive (Mellon 2008).

Given the climate of debate over agricultural biotechnology, the goal of risk communication shifted from the approach typical in public health contexts to one in which the goal is *two-way communication*. Not only does the public become better informed about risks of biotechnology, the government and technology developers become better informed about the views of the public (Priest 1995; Covello and Sandman 2001). Methods for actually undertaking such a communications effort were proposed especially in Europe, where GMOs elicited particularly effective political resistance (Rowe and Frewer 2000). A detailed discussion of the successes and failures of these efforts is out of place in the present context. Suffice it to say, not everyone is a fan. Critics have suggested that the primary result of such efforts has been to subvert more traditional forms of political debate over new technology. In place of true debate, these risk communication and public engagement efforts become converted into efforts at tempering public resistance to emerging technologies (Irwin 2006; Rayner 2007). This is what people who study risk communication in the nuclear power/biotechnology line of descent call *shilling for industry*.

Where does nanotechnology fit with respect to these two traditions in risk communication? One answer places it squarely in a line of descent from nuclear power and biotechnology. In both the United States and Europe, funding for social science and ethics research on emerging nanotechnologies was frequently justified by a desire to avoid the debacle experienced with GMOs (Berubi 2006). Scholars who undertook studies of nanotechnology often did so on the basis of prior experience in studying resistance to GMOs (Gaskell 2008; Macnaughten 2008). This would suggest that someone developing an ethics of risk communication for nanotechnology would be well advised to adopt a certain amount of skepticism about what proponents of nanotechnology say about their technology and also to think broadly, rather than narrowly, about what topics a risk communication effort should encompass. The fact that some prominent social critics of nanotechnology are also graduates of the GMO wars would give additional credence to this warning (ETC Group 2002, 2004; Miller and Senjen 2008).

It is also the answer that has been behind a substantial amount of the social science research that has been conducted in connection with nanotechnology. Anticipatory governance, for example, was hailed as an attempt to, among other things, make developers of new technology more sensitive to features that would trigger public resistance and concern at the earliest possible point in the innovation process, and certainly well before key decisions in product development were made (Barben et al. 2007). This, I hasten to add, is *not* to say that anticipatory governance is equivalent to risk communication, or that either is inherently or irrevocably committed to the task of easing the road for public acceptance of nanotechnology. The point here is simple. To the extent that nanotechnology risk communication is situated within a tradition of scholarship and practice that understands the task of risk communication to derive from the history of emerging technologies like nuclear power and GMOs, the ethics of risk communication will tend to be more concerned about shilling the industry than scaring the pants off of people. Or to put it less gracefully, perhaps it is *because* they are shilling for industry that risk communicators will be worried about scaring the pants off of people. One does not bite the hand that feeds one.

This is not the only answer that is available, however. An alternative would be to step back from fascination with nanotechnology as such and inquire more carefully about the specific technical applications being developed. Many of the signature applications of nanotechnology lie in the field of medicine, where even biotechnologies were never particularly controversial (Bauer 2002). What is more, the term *nanotechnology* has been stretched to encompass many chemical applications that were well underway long before key funding initiatives to promote nanotechnology were initiated. In many cases, these applications do not involve features such as convergence of electronic, biological, and chemical capabilities, though they often do exploit chemical reactivity or bonding properties that emerge at the nanoscale (Maynard 2007). It is at least an open question as to whether a development that would clearly have occurred irrespective of the nanotechnology hype, and would almost certainly not ever have been referred to as an application of nanotechnology, is usefully included in a risk communication effort designed to gauge the public's response to a genuinely novel technological development. This is not to imply that these applications are without risk. Rather, the question is whether they pose the kind of risk that people genuinely *want* to be consulted about in the course of regulation and governance (Thompson 2010).

Given this alternative way of thinking, "scaring the pants off of people" might happen simply *because* an organized effort of risk communication for nanotechnology is launched. As I will discuss below, just mentioning risk is a signal that people should be on their guard. If what we are talking about are new technologies that might pose hazards, but hazards that toxicologists, engineers, and other specialists in risk analysis know how to evaluate, it is questionable as to whether the kind of efforts envisioned in the wake of nuclear power and GMOs are justified by the magnitude of the risk or by limitations in the ethics of the scientific community. This is a big "if." Some social science commentators on nanotechnology would clearly reject the antecedent of that conditional (Cobb and Macoubrie 2004; Nordmann and Schwarz 2010). At the same time, if the lure of big science *is* driving a number of developments, only some of which happen to have anything to do with risk, then risk communication activities aimed to promote any kind of measured decision-making process on the acceptability of nanotechnology may *deserve* to be classified as shilling for industry. Then we are right back into the nuclear power/GMO story all over again. I am sorry if this less than totally succinct statement of the ethical task has failed to clarify things for the budding risk communicator.

But as warned, there are also matters that have complicated the ethics of risk communication markedly, even presuming that one has decided what course to steer on matters of substance. These matters have little to do with nanotechnology as such, but a great deal to do with the intermingling of ethics and risk. At the outset, one might note the way that predominant approaches in risk analysis and risk management have tended to be closely associated with the utilitarian style of decision making and regulatory policy. This is an approach that interprets the general problem of managing risks from technology as one of striking the optimal balance between benefit and risk. This approach fits well with the public health tradition of risk communication, where scientists know that current behaviors have produced a risk/benefit ratio that is uncontroversially suboptimal. But generalizing this way of thinking about risk to all areas of policy is a mistake. There are many commentaries on this problem in the literature (Schulze and Kneese 1981; Rayner and Cantor 1987; Brunk et al. 1991). A clear alternative to the utilitarian perspective can be found in the criterion applied to risk decisions in the case of medical treatments and human subjects research: Risks are acceptable only if they are actually accepted by the people who must bear them as determined by procedures of informed consent. Here the point of a risk communication is to empower the people who are being asked to bear the risks. It is *not* to convince them that a particular choice is the rational one. The ethical rationale for informed consent can be traced to Kantian or deontological views in ethics, the main competitor to a utilitarian approach (Donagan 1977; Macklin 1992).

The overall ethical framework—utilitarian or Kantian—in which a given risk problem is conceptualized can thus play a significant role in shaping the way one would develop an appropriate communication effort. In cases where informed consent has been clearly identified as the appropriate standard, as in developing protocols for the use of human subjects in research contexts, a communication tool should clearly try to minimize its persuasive component. It would be unethical to use the kind of behavior change tactics that are commonplace in the public health tradition. At the same time, some of the most widely studied problem areas involve educational and persuasion efforts where the attempt to achieve behavior change would appear to be ethically well justified. The role of risk communication in policy making for new technology is far less clear and would reopen points that have already been covered above. Is risk communication supposed to make people think like risk experts? Is it to enable the accomplishment of those social goals that would be endorsed by utilitarian or Kantian ethical theories, respectively? Even if one remains agnostic about the answers to such questions, specialists in risk communication would be well advised to study the relationship between assumed criteria for ethical decision making and the functions of risk communication more carefully. At a minimum, simply presuming that the least ethically complex public health cases of risk communication are prototypical is unjustified.

There are also a number of well-known foibles that trip up efforts at risk communication, and some of them have a lot to do with ethics (s.). One of them has been widely discussed within risk analysis as a multidisciplinary field of study. Using science and engineering to inform judgments of risk demands a certain separation between the technical specification of hazard and exposure and the admittedly value-driven activity of deciding what to do (risk management).

But how do we specify this certain separation? Lots of ink has been spilled over whether the technical aspects of risk assessment can be value free. Although there are people who say that they can be and should remain so (Cross 1998), the short answer is that they are wrong (Thompson and Dean 1996; Cranor 1997; Thompson 2003). Some progress toward getting clear on what is at stake here can be made by drawing a distinction between social values, which express preferences toward specific social outcomes, and epistemic values, which specify the norms for scientific inquiry (Ben-Ari and Or-Chen 2009).

This rather technical problem rears its head in risk communication when scientists and engineers expect risk communicators to adhere rigidly to their rather peculiar conceptualization of risk as a mathematical relationship between hazard and exposure when undertaking a communication activity. In ordinary language, the value dimension of risk comes to the fore. In ordinary nontechnical discourse, risk is strongly associated with agency, with the fact that someone—the person imposing or bearing the risk—is *doing* something (Tulloch and Lupton 2003; Hamilton et al. 2008). From this starting point, it is easy to move to the suggestion that doing something different is the responsible thing to do, and then we are in the territory of ethics. Telling people that everything is hunky dory and that there is really no reason for doing anything is very tricky. If everything is fine, why are you standing there trying to reassure me so assiduously? The inherently moralized context of risk discourse thus leads risk communications that are (perhaps reasonably) intended to dampen fires and quiet fears into some very slippery rhetorical turf (Katz 2001).

Perhaps even more intractable difficulties have been introduced by the emergence of a sociology of risk in the two decades since Ulrich Beck has been translated into English. Beck's book *Risk Society* proposed the thesis that people in contemporary societies are becoming obsessed with risk, and that debates over risk are replacing earlier debates over class divisions (Beck 1992). There has since been a veritable explosion of work in the sociology of risk. Some of it is incredibly silly. One often reliable indicator of silliness is the use of the word *reflexive* when reflection or conscious, reflective deliberation is meant. Reflexivity, in contrast, denotes a class of actions or behaviors that operate on reflex—that is, *without* conscious, reflective deliberation. Because ethics of either an s. or p. sort aims for conscious, reflective deliberation, in place of reflexive response, this point is not trivial in the present context. Here is a point on which a rather bizarre convention in social theory has had obfuscatory consequences. Beck is quite aware of this problem (Beck 1994), but his protestation has done little to stem the tide.

Of more general importance to risk communication is the way that sociological studies of risk have plunged risk communication into an inherently normative debate over the place of risk in social theory. Beck's promotion of risk irks theorists who see class relations as the fundamental organizational principle in industrial societies, and they have both theoretical and normative (ethical) reasons for trying to beat Beck's views down. For example, if you are "pro-poor," you might find an emphasis on the way all of us are at risk in postmodernity to be a rather retrograde form of politics, one that threatens to undercut the goals of distributive justice. Although this example oversimplifies a complex debate, it illustrates how conceptually large themes in social and political theory are capable of subsuming what might have once been thought to be much narrower attempts to promote behavioral change for public health or study emerging technology. Suddenly, risk communication efforts of all kinds have the potential to be politicized in novel ways. Of course, some readers of this essay on the ethics of nanotechnology risk communication have almost certainly come to the topic *because* they assume that "grand narrative" issues like this one are at stake. Nanotechnology is but a pawn in the larger game of history, and we play our moves based on our allegiances in that larger game.

Ethical and theoretical viewpoints on the larger game may circumscribe efforts at nanotechnology risk communication. This cannot help but influence whether one thinks that scaring the pants off of people or, alternatively, shilling for industry, are perfectly appropriate things to do. But being able to take a stance that adopts a certain critical distance from the larger game is, nonetheless, what ethics (s.) has always been about (Thompson 2010). And that is what scholars of risk communication should do.

REFERENCES

Barben, D., E. Fisher, C. Selin, and D.H. Guston. 2007. Anticipatory governance of nanotechnology: foresight, engagement, and integration. In *The Handbook of Science and Technology Studies*. Eds. E.J. Hackett, O. Amsterdamska, M. Lynch, and J. Wajcman. Cambridge, MA: MIT Press, 979–1000.

Barkan, S.E. 1970. Strategic, tactical and organizational dilemmas of the protest movement against nuclear power. *Social Problems* 27: 19–37.

Bauer, M. 2002. Controversial medical and agri-food biotechnology: a cultivation analysis. *Public Understanding of Science* 11: 93–111.
Beck, U. 1992. *Risk Society: Towards a New Modernity.* London: Sage.
Beck, U. 1994. Self-dissolution and self-endangerment of industrial society: what does this mean? In *Reflexive Modernization: Politics, Tradition and Aesthetics in the Modern Social Order.* Eds. U. Beck, A. Giddens, and S. Lash. Stanford, CA: Stanford University Press, 174–183.
Ben-Ari, A., and K. Or-Chen. 2009. Integrating competing conceptions of risk: a call for future direction of research, *Journal of Risk Research* 12: 865–877.
Berubi, D. 2006. *Nano-Hype: The Truth behind the Nanotechnology Buzz.* Amherst, NY: Prometheus Books.
Brunk, C., L. Haworth, and B. Lee. 1991. *Value Assumptions in Risk Assessment: A Case Study of the Alachlor Controversy.* Waterloo, ON: Wilfrid Laurier University Press.
Cates, W. Jr., D.A. Grimes, H.W. Ory, and C.W. Tyler, Jr. 1977. Publicity and public health: the elimination of IUD-related abortion deaths. *Family Planning Perspectives* 9: 138–140.
Cho, H., and K. Witte. 2005. Managing fear in public health campaigns: a theory-based formative evaluation process. *Health Promotion Practice* 6: 482–490.
Cobb, M.D., and J. Macoubrie. 2004. Public perceptions about nanotechnology: risks, benefits and trust. *Journal of Nanoparticle Research* 6(4): 395–405.
Cook, E. 1982. The role of history in acceptance of nuclear power. *Social Science Quarterly* 63: 3–16.
Covello, V., and P.M. Sandman. 2001. Risk communication: evolution and revolution. In *Solutions to an Environment in Peril.* Ed. Anthony Wolbarst. Baltimore, MD: Johns Hopkins University Press, 164–178.
Cranor, C.F. 1997. The normative nature of risk assessment: features and possibilities. *Risk: Health, Safety and Environment* 8: 123–136.
Cranor, C.F. 2006. *Toxic Torts: Science, Law and the Possibility of Justice.* New York: Cambridge University Press.
Cross, F.B. 1998. Facts and values in risk assessment. *Reliability Engineering and System Safety* 59: 27–40.
Donagan, A. 1977. Informed consent in therapy and experimentation. *Journal of Medicine and Philosophy* 2: 307–329.
ETC-Group. 2002. *The Big Down: From Genomes to Atoms.* Montreal, Canada: ETC Group.
ETC-Group. 2004. *Down on the Farm: The Impact of Nano-scale Technologies on Food and Agriculture.* Montreal, Canada: ETC Group.
Gaskell, G. 2008. Lessons from the bio-decade: a social science perspective. In *What Can Nanotechnology Learn from Biotechnology: Social and Ethical Lessons for Nanoscience from the Debate over Agrifood Biotechnology and GMOs.* Eds. K. David and P.B. Thompson. Burlington, MA: Academic Press, 237–259.
Hamilton, C., S. Adolphs, and B. Nerlich. 2008. The meanings of "risk": a view from Corpus Linguistics. *Discourse and Society* 18: 163–181.

Irwin, A. 2006. The politics of talk: coming to terms with the "new" scientific governance. *Social Studies of Science* 36(2): 299–320.

Katz, S.B. 2001. Language and persuasion in biotechnology communication with the public: how to not say what you're not going to not say and not say it. *AgBioForum* 4: 93–97.

Lapinski, M.K., and F.J. Boster. 2001. Modeling the ego-defensive function of attitudes. *Communication Monographs* 68: 314–324.

Lewis, H.W. 1990. *Technological Risk.* New York: W.W. Norton.

Lovins, A. 1977. *Soft Energy Paths: Toward a Durable Peace.* San Francisco: Friends of the Earth International.

Macklin, R. 1992. Universality of the Nuremberg Code. In *The Nazi Doctors and the Nuremberg Code.* Eds. G.J. Annas and M.A. Grodin. New York: Oxford University Press, 240–257.

Macnaughten, P. 2008. From bio to nano: learning the lessons, interogating the comparisons. In *What Can Nanotechnology Learn from Biotechnology: Social and Ethical Lessons for Nanoscience from the Debate over Agrifood Biotechnology and GMOs.* Eds. K. David and P.B. Thompson. Burlington, MA: Academic Press, 107–124.

Maynard, A.D. 2007. Nanotechnology: the next big thing, or much ado about nothing? *Annals of Occupational Hygene* 51: 1–12.

McHughen, A. 2008. Learning from mistakes: missteps in public acceptance issues with GMOs. In *What Can Nanotechnology Learn from Biotechnology: Social and Ethical Lessons for Nanoscience from the Debate over Agrifood Biotechnology and GMOs.* Eds. K. David and P.B. Thompson. Burlington, MA: Academic Press, 33–54.

Mellon, M. 2008. A view from the advocacy community. In *What Can Nanotechnology Learn from Biotechnology: Social and Ethical Lessons for Nanoscience from the Debate over Agrifood Biotechnology and GMOs.* Eds. K. David and P.B. Thompson. Burlington, MA: Academic Press, 81–87.

Michaels, D. 2008. *Doubt Is Their Product: How Industry's Assault on Science Threatens Your Health.* New York: Oxford University Press.

Miller, G., and R. Senjen. 2008. "Out of the laboratory and on to our plate: nanotechnology in food and agriculture." Friends of the Earth, Washington, DC. http://foe.org/out-laboratory-and-our-plates

Miller, H.I., and G. Conko. 2001. Precaution without principle. *Nature Biotechnology* 19: 302–303.

Nordmann, A., and A. Schwarz. 2010. The lure of the "yes": the seductive power of technoscience. *Governing Future Technologies: Sociology of the Sciences Yearbook* 27: 255–277,

Priest, S. 1995. Information equity, public understanding of science, and the biotechnology debate. *Journal of Communication* 45: 39–54.

Priest, S. 2000. *A Grain of Truth: The Media, the Public and Biotechnology.* Lanham, MD: Rowman and Littlefield.

Rayner, S. 2007. The rise of risk and the decline of politics. *Environmental Hazards* 7(2): 165–172.

Rayner, S. and R. Cantor. 1987. How fair is safe enough? The cultural approach to societal technology choice. *Risk Analysis* 7: 3–9.

Rowe, G., and L.J. Frewer. 2000. Public participation methods: a framework for evaluation. *Science, Technology, and Human Values* 25: 3–29.

Schulze, W.D., and A.V. Kneese. 1981. Risk in benefit-cost analysis. *Risk Analysis* 1: 81–88.

Slovic, P. 1986. Informing and educating the public about risk. *Risk Analysis* 6: 403–415.

Slovic, P. 1987. Perception of risk. *Science* 236: 280–285.

Starr, C., and C. Whipple. 1980. Risks of risk decisions. *Science* 208: 1114–1119.

Thompson, P.B. 2003. Value judgments and risk comparisons: the case of genetically engineered crops. *Plant Physiology* 132: 10–16.

Thompson, P.B. 2010. Agrifood nanotechnology: is this anything? In *Understanding Nanotechnology: Philosophy, Policy and Publics.* Eds. U. Fiedeler, C. Coenen, S.R. Davies, and A. Ferrari. Heidelberg, Germany: Akademische Verlagsgesellshchaft AKA GmbH, 157–170.

Thompson, P.B., and W.E. Dean. 1996. Competing conceptions of risk. *Risk: Health, Safety and Environment* 7: 361–384.

Thompson, P.B., and W. Hannah. 2008. Food and agricultural biotechnology: a summary and analysis of ethical concerns. *Advances in Biochemical Engineering and Biotechnology* 111: 229–264.

Tulloch, J., and D. Lupton. 2003. *Risk and Everyday Life.* London: Sage.

Index

A

Agenda building, 153
Amplification stations, 28, 65
Anti-nanotechnology protestors, 66
Asynchronous communication, 108
Attentive public, 156

B

Blank slate, 107
Booster coverage, 148
Breakthrough research, 144
Bureaucracies, 114

C

Childhood diseases, 43
Citizen panel interviews, 87
Clean Water Act, 36
Climate change, 27
  global, 3, 86
  polluting technologies and, 3
  societal reactions, 28
  technology and, 1
  Web sites, 145
Cloning
  assertion about, 6
  emotional reaction, 7
  ethical concerns, 5, 32
  self-identity and, 121
Collective behavior, 47
Communication, *see also* Risk communication
  asynchronous communication, 108
  crisis, 39, 169
  ethics, 171–180
  professional principles, 33–37
  research, 124
  synchronous, 108
  two-way, 175
Consequential issue analysis, 6
Crisis communication, 39, 169
Cultural resonance, 82
Cultural theories of risk, 48
Cultures, nanotechnology risk and, 113–141
  bureaucracies, 114
  farm managers, 116
  international trade agreements, 115
  nanobiotechnology, 125
  nanotechnology governance, 126–131
  orphan diseases, 119
  persuasion research, 124–125
  policy think tanks, 116
  Prey effect, 135
  psychometric paradigm, 120
  public failure model, 130
  public policies, 132–136
  risk averse individuals, 119
  risk regulation, 118
  role of advocacy in democracy, 116–119
  self-identity, 121
  self-regulation, 115
  social, cultural, and psychological influences, 120–124
  stem cell research, 121
  technocentric model, 120
  technologies of humility, 131
  tradition within communication research, 124
  unknown business liabilities, 115

D

Deliberative democracy, 83
Democracy
  deliberative, 83
  "public sphere," 84
  representative, 85
  role of advocacy in, 116–119
Democratic society, risk communication in, 1–19
  climate change, technology and, 1
  cloning, assertion about, 6
  consequential issue analysis, 6
  "father of nanotechnology," 13
  microcrystalline materials, 13

microfabrication, 15
microscience, 11
molecular engineering, 12
nanoscale, 2
public engagement, 9
rights-based perspective, 6
risk societies, 4
risk and technology, 3–6
social risk, 5
technology and society, 6–9
upstream engagement, 9
utilitarian, issue analysis, 6
where nano came from, 9–16

E

Emotional reactions, 5
EPA, *see* U.S. Environmental Protection Agency
Ethics, 171–180
Event reporting, 158–159
Exclusive scoop, 147
Expert opinions versus public opinion, 74

F

Farm managers, 116
"Father of nanotechnology," 13
Federal Insecticide, Fungicide, and Rodenticide Act, 129
Food packaging, nanotechnology-based, 2
Freedom of Information Act, 36
"Functionalist" sociology, 146
Future challenges, 167–183
Citizens' Forums, 171
crisis communication, 169
ethics of risk communication, 171–180
genetically modified organisms, 174
hazards, 167
lay citizens, 170
nanobiotechnology, 167
newsroom power, shrinking, 170
nuclear power, 175
organizational principle, 180
philosophers, 172
pro-poor, 180
shilling for industry, 173, 175
social amplification of risk, 168
social theory, bizarre convention in, 180
synthetic biology techniques, 167
Three Mile Island accident, 174
two-way communication, 175
uncertainty, 169

G

Gene therapy, 6
Genetically modified (GM) food, 47, 91
Genetically modified organisms (GMOs), 174
GMOs, *see* Genetically modified organisms

I

Industrial revolution, 148
Information, dissemination of about new technologies, 143–166
agenda building, 153
attentive public, 156
balanced view, 148
booster coverage, 148
breakthrough research, 144
conflict theory, 148
exclusive scoop, 147
"functionalist" sociology, 146
industrial revolution, 148
marketplace of ideas, 144
mass media reporting, 157–164
conclusions about overall media coverage, 163–164
event reporting, 158–159
information sources, 162–163
nanotechnology risk articles, 158
regulatory issues, 161–162
study, 157–158
top generators of risk news, 160–161
types of risk coverage, 159–160
media effects theory in Internet age, 150–155

media messages, knee-jerk reaction to, 153
media's role in risk societies, 146–150
museums and science centers, 155–156
NanoDays, 156
news media, 143
perception of risk, 150
public sphere discussion, 143
social amplification studies, 147
two-sided issue, 145
"watchdog" advocacy groups, 147
wire service coverage, 161

L

Lay citizens, 170
Legislation
Clean Water Act, 36
Federal Insecticide, Fungicide, and Rodenticide Act, 129
Freedom of Information Act, 36
National Environmental Policy Act, 36
Right to Know Act, 36
Toxic Release Inventory, 36
Toxic Substances Control Act, 113
Lessons and future challenges, 167–183
Citizens' Forums, 171
crisis communication, 169
ethics of risk communication, 171–180
genetically modified organisms, 174
hazards, 167
lay citizens, 170
nanobiotechnology, 167
newsroom power, shrinking, 170
nuclear power, 175
organizational principle, 180
philosophers, 172
pro-poor, 180
shilling for industry, 173, 175
social amplification of risk, 168
social theory, bizarre convention in, 180
synthetic biology techniques, 167
Three Mile Island accident, 174
two-way communication, 175
uncertainty, 169

M

Magic Nano, 23
Marketplace of ideas, 26, 40, 144
Mass media reporting, 157–164
conclusions about overall media coverage, 163–164
event reporting, 158–159
information sources, 162–163
nanotechnology risk articles, 158
regulatory issues, 161–162
study, 157–158
top generators of risk news, 160–161
types of risk coverage, 159–160
Microcrystalline materials, 13
Microscience, 11
Model
deficit model, 24, 26
marketplace of ideas model, 26
public failure, 130
technocentric model, 120
Molecular engineering, 12

N

Nanobiotechnology, 5, 125
Nanoscale, 2
Nanotechnology
application of, 177
central idea of, 14
communication about, 31
cultural resonance and, 68–70
definition of, 1
envisioning, 22
food packaging, 2
framing of risk messages, 45
gene therapy, 6
global development, 115
governance, 126
materials science, 2
public attitudes toward, 71–75
expert opinions versus public opinion, 74
remaining challenges, 74–76

types of attitudes and cognitions that matter most, 72–74
where public attitudes come from, 71–72
research, 117
synonym, 12
Nanotechnology, introduction of to the public, 21–38
amplification stations, 28
climate change, 27
deficit model, 24, 26
democratic theory, 27
early adopters, 31
emerging public perception, 26–29
emotional reactions, 5
imagining the nanoscale, 22–25
Magic Nano, 23
marketplace of ideas model, 26
nanobiotechnology, 5
popular concerns, 23
professional principles of risk communication, 33–37
risk communication for twenty-first century democracies, 29–33
social amplification of risk, 28
upstream public engagement, 27
Nanotechnology risk, 113–141
bureaucracies, 114
farm managers, 116
international trade agreements, 115
nanobiotechnology, 125
nanotechnology governance, 126–131
orphan diseases, 119
persuasion research, 124–125
policy think tanks, 116
Prey effect, 135
psychometric paradigm, 120
public failure model, 130
public policies, 132–136
risk averse individuals, 119
risk regulation, 118
role of advocacy in democracy, 116–119
self-identity, 121
self-regulation, 115
social, cultural, and psychological influences, 120–124
stem cell research, 121
technocentric model, 120
technologies of humility, 131
tradition within communication research, 124
unknown business liabilities, 115
National Environmental Policy Act, 36

O

One-shot research designs, 64
Opinion leadership, 48
Organizational principle, 180
Orphan diseases, 191

P

Prey effect, 135
Psychometric paradigm, 120
Public consultation practices, 89
Public failure model, 130
Public opinion, public perception, and public understanding, 57–79
amplification stations, 65
anti-nanotechnology protestors, 66
attentive public, 61
GM food story revisited, 58–61
implications for risk communication research and practice, 65–68
nanotechnology and cultural resonance, 68–70
one-shot research designs, 64
opinion studies and their implications, 61–65
polling, 63
public attitudes toward nanotechnology, 71–75
expert opinions versus public opinion, 74
remaining challenges, 74–76
types of attitudes and cognitions that matter most, 72–74
where public attitudes come from, 71–72
recombinant bovine somatotropin, 58
socioeconomic status, 75

"Public sphere" democracy, 84

R

Representative democracy, 85
Research designs, one-shot, 64
Right to Know Act, 36
Rights-based perspective, 6
Risk communication
best practices, 46
autism, 43
challenges, 40–44
childhood diseases, 43
collective behavior, 47
crisis communication, 39
cultural theories of risk, 48
deficit model, 39
diffusion theory, 46
framing of risk messages, 45
genetically modified food, 47
goals of risk communication, 44–46
mad cow disease, 43
marketplace of ideas philosophy, 40
opinion leadership, 48
recommendations for improving risk communication, 51–55
social amplification theory, 50
social theories of risk, 46–51
spiral of silence, 47
Risk societies, 4

S

SES, *see* Socioeconomic status
Shilling for industry, 173, 175
Social amplification of risk, 28
Social risk, 5
Social theories of risk, 46
Socioeconomic status (SES), 75
Spiral of silence, 47
Stakeholders, nanotechnology risk and, 113–141
bureaucracies, 114
farm managers, 116
international trade agreements, 115
nanobiotechnology, 125
nanotechnology governance, 126–131
orphan diseases, 119
persuasion research, 124–125
policy think tanks, 116
Prey effect, 135
psychometric paradigm, 120
public failure model, 130
public policies, 132–136
risk averse individuals, 119
risk regulation, 118
role of advocacy in democracy, 116–119
self-identity, 121
self-regulation, 115
social, cultural, and psychological influences, 120–124
stem cell research, 121
technocentric model, 120
technologies of humility, 131
tradition within communication research, 124
unknown business liabilities, 115
Stem cell research, 121

T

Technocentric model, 120
Technocracy, 86
Technological literacy, 86
Technology, public demands, 81–112
amplified risks, 92
asynchronous communication, 108
attenuated risks, 92
blank slate, 107
challenges of risk society, 85–89
citizen panel interviews, 87
consensus conferences, 95–96
converging technologies, 95
cultural resonance, 82
deliberative democracy, 83
drawing room discussions, 84
genetically modified foods, 91
market researchers, 89
nanotechnology, risk, and society, 89–93
National Citizens' Technology Forum, 95–96
odd bedfellows, 89
public consultation practices, 89

"public sphere" democracy, 84
representative democracy, 85
synchronous communication, 108
technocracy, 86
technological literacy and democracy, 82–85
technology conference, lessons learned, 93
template of hopes and fears, 88
template of technology-related concerns, 87
upstream public management, 81
Theory
conflict, 148
democratic, 27
diffusion, 46
media effects (in Internet age), 150–155
social, bizarre convention in, 180
social amplification, 50
Theory and practice, risk communication in, 39–56
autism, 43
childhood diseases, 43
collective behavior, 47
crisis communication, 39
cultural theories of risk, 48
deficit model, 39
diffusion theory, 46
framing of risk messages, 45
genetically modified food, 47
goals of risk communication, 44–46
mad cow disease, 43
marketplace of ideas philosophy, 40
opinion leadership, 48
recommendations for improving risk communication, 51–55
risk communication
best practices, 46
challenges, 40–44
social amplification theory, 50
social theories of risk, 46–51
spiral of silence, 47
Toxic Release Inventory, 36
Toxic Substances Control Act, 113

U

Uncertainty, 169
Unknown business liabilities, 115
Upstream engagement, 9, 81
U.S. Environmental Protection Agency (EPA), 113
Utilitarian, issue analysis, 6

W

Wire service coverage, 161

For Product Safety Concerns and Information please contact our EU representative GPSR@taylorandfrancis.com Taylor & Francis Verlag GmbH, Kaufingerstraße 24, 80331 München, Germany